Great Lakes
Heroes & Villains

WAYNE LOUIS KADAR

Avery Color Studios, Inc.
Gwinn, Michigan

ISBN-13: 978-1-892384-51-5
ISBN-10: 1-892384-51-5

Library of Congress Control Number: 2009922381

First Edition–2009

10 9 8 7 6 5 4 3 2 1

Published by Avery Color Studios, Inc.
Gwinn, Michigan 49841

Cover photos: Author's Collection

This book is dedicated to the courageous men and women of the United States and Canadian Coast Guard. All too often their deeds are unsung. They risk their lives each day to protect lives and property of recreational boaters, commercial vessels and in today's society they are also asked to protect the major waterways of the United States and Canada from threats from outside the country.

We recreational boaters complain about gusty winds, thunderstorms and high seas that keep us tied to the docks, but it is these conditions that the Coast Guard train in, for these are the conditions when they are called out.

The book is also dedicated to the recreational and commercial boaters who go to the aid of boaters in need of assistance. When a call is made on the radio that a boat is disabled, someone is lost in the fog or there is an onboard injury, recreational boaters are known to pull fishing lines, alter their course when cruising and take skiers ashore to go to help the boater in need.

Table of Contents

Heroes Of The Great Lakes

Villains Of The Great Lakes

Introduction To The Heroes Of The Great Lakes

Throughout the long and varied history of the Great Lakes region, men and women of the Life-Saving Service, the Lighthouse Establishment and the United States and Canadian Coast Guard have risked their lives for persons or property in peril on the lakes. Each day these heroes stand prepared to leave the dock in conditions that chase all others to shelter. They use their skills to save those who find themselves in conditions beyond their control.

While those in the military are not the only heroes, sometimes men and women working on the lakes or shore-side occupations find themselves thrust into a position where it is their actions that can save the lives of those in jeopardy. Mostly untrained and running on adrenaline these civilians risk all to save the lives of strangers.

The following are just a few of the heroes, both civilian and military, who have distinguished themselves as heroes of the Great Lakes.

History Of The United States Life-Saving Service And Coast Guard

The first organized efforts in the United States to assist sailors in peril came in 1786 just ten years after the country was founded. The Massachusetts Humane Society, a civilian organization, erected huts of refuge along the Massachusetts Atlantic Ocean shore. The huts provided shelter for sailors whose ships had foundered or grounded and were fortunate enough to make it to shore. All too often sailors made it off their sinking ship only to die on shore from exposure or lack of nourishment.

The congress of the United States in 1790 authorized the construction of ten vessels to enforce the new country's tariff and trade laws. The ships were also charged with preventing the smuggling of contraband goods into the new country. This branch of the new government was named the Revenue Marine Service and later changed to the Revenue Cutter Service.

Over the next several decades the Massachusetts Humane Society established lifeboat stations manned by volunteers and several more huts of refuge. In 1847 the United States entered into the life-saving business when congress appropriated funds to equip the lighthouses along the coast with lifeboats and equipment so they could render assistance to ships in trouble. A few years later congress approved $10,000.00 for "...providing surf boats, rockets, carronades and other necessary apparatus for the better preservation of life and property from shipwreck on the coast of New Jersey."

It wasn't until 1854 that the Great Lakes were

awarded funding for life-saving efforts. Forty-two thousand, five hundred dollars was appropriated for fourteen additional stations in New Jersey, eleven on the coast of Long Island and twenty-three lifeboats for use on Lake Michigan.

The life-saving stations continued to grow in number but volunteers manned them and there was not any organization to the efforts until 1871 when congress provided $200,000 and authorized the Secretary of Treasury to create the Revenue Marine Service to employ surfmen at the stations.

Sumner Kimball was appointed to develop the new agency. He inspected stations and found most of them to be in deplorable condition; buildings in varying states of disrepair, some keepers, too aged or ill to function in their position, and equipment missing or in poor condition. Mr. Kimball set out to correct these shortcomings and improve the organization and function of the service.

Under the guidance of Sumner Kimball, Surfmen were carefully selected for their ability to do the job. New rescue equipment was purchased and he established qualifications for administrative positions.

There was an Inspector of Lifesaving stations over all of the districts. Reporting to him were the superintendents who were in charge of each district.

To be appointed a District Superintendent a person had to pass an examination, live in the district, had to be familiar with the coast and surf conditions of the district, and had to be familiar with lifeboats and other lifesaving apparatus.

Under the direction of the District Superintendent were the keepers of each station. To be qualified as a keeper one had to: "...be the best that can be obtained from the athletic race of beachmen, a master of boat craft, and the art of surfing, and skilled in wreck operations." Once the keeper of a station was selected he would then select his own surfmen.

By 1877 the importance of commercial shipping on the Great Lakes was acknowledged. Three new districts were established, the Ninth District encompassed the stations on the American coasts of Lakes Ontario and Erie. The stations in the Tenth District covered Lakes Huron and Superior, with the Lake Michigan stations making up the Eleventh District.

The following are the 1899 locations of the life-saving stations of the three Great Lakes districts.

The Ninth District - Lakes Ontario and Erie

Oswego, New York, Lake Ontario, north side of the entrance of Oswego Harbor, established 1876.

Fort Niagara, New York, Lake Ontario, east side entrance to the Niagara River. Established 1892.

Charlotte, New York, Lake Ontario, east side entrance of Charlotte Harbor. Established 1876.

Buffalo, New York, Lake Erie, south side of the entrance of Buffalo Harbor. Established 1877.

Erie, Pennsylvania, Lake Erie, north side entrance. Established 1894.

Cleveland, Ohio, Lake Erie, west side entrance of Cleveland Harbor. Established 1876.

Ashtabula, Ohio, Lake Erie, west side of the harbor entrance. Established 1876.

Fairport, Ohio, Lake Erie, west side of the Fairport Harbor. Established 1876.

Point Marblehead, Ohio, Lake Erie, on Marblehead Island. Established 1876.

The Tenth District - Lakes Huron and Michigan

Sand Beach, Michigan, Lake Huron, inside Sand Beach Harbor of Refuge. Established 1881.

Point Aux Barques, Michigan, Lake Huron, near lighthouse. Established 1876.

Grindstone City, Michigan, Lake Huron, two miles northwest of Grindstone City. Established 1881.

Ottawa Point, Michigan, Lake Huron, located near the lighthouse. Established 1876.

Sturgeon Point, Michigan, Lake Huron, near the lighthouse. Established 1876.

Thunder Bay Island, Michigan, Lake Huron, near lighthouse. Established 1876.

Middle Island, Michigan, Lake Huron, north end of Middle Island. Established 1881.

Hammond Bay, Michigan, Lake Huron, near the Forty Mile Point Lighthouse. Established 1876.

Bois Blanc, Michigan, Lake Huron, midway on the east side of the island. Established1891.

Vermillion Point, Michigan, Lake Superior, ten miles west of Whitefish Point. Established 1876.

Crisp, Michigan, Lake Superior, eighteen miles west of Whitefish Point. Established 1876.

Two-Hearted River, Michigan, Lake Superior, near the mouth of the Two-Hearted River. Established 1876.

Muskellunge, Michigan, Lake Superior, near the mouth of the Sucker River. Established 1876.

Marquette, Michigan, Lake Superior, near the lighthouse. Established 1891.

Ship Canal, Michigan, Lake Superior, near the mouth of Portage Lake and the Lake Superior ship canal. Established 1884.

Duluth, Minnesota, Lake Superior, on Minnesota Point. Established 1895.

Eleventh District - Lake Michigan

Beaver Island, Michigan, Lake Michigan, near the lighthouse. Established 1876.

North Manitou Island, Lake Michigan, near Pickard's Wharf. Established 1876.

Point Betsey, Michigan, Lake Michigan, near the Point Betsey lighthouse. Established 1876.

Frankfort, Michigan, Lake Michigan, on the south side of the harbor entrance. Established 1887.

Manistee, Michigan, Lake Michigan, north side of the harbor entrance. Established 1879.

Grand Point au Sable, Michigan, Lake Michigan, one mile south of the lighthouse. Established 1876.

Ludington, Michigan, Lake Michigan, north side of the entrance of the harbor. Established 1879.

Pentwater, Michigan, Lake Michigan, north side of the harbor entrance. Established 1879.

White River, Michigan, Lake Michigan, north side of the entrance to White Lake. Established 1887.

Muskegon, Michigan, Lake Michigan, north side of the harbor entrance. Established 1879.

Grand Haven, Michigan, Lake Michigan, north side of harbor entrance. Established 1876.

Holland, Michigan, Lake Michigan, south side of the harbor. Established 1886.

South Haven, Michigan, Lake Michigan, south side of the harbor entrance. Established 1887.

St. Joseph, Michigan, Lake Michigan, in the harbor on the north side. Established 1887.

Michigan City, Indiana, Lake Michigan, on the east side of the entrance to the harbor. Established 1889.

South Chicago, Illinois, Lake Michigan, north side of the entrance to Calumet Harbor. Established 1890.

Chicago, Illinois, Lake Michigan, seven miles south and east of the Chicago River Lighthouse. Established 1893.

Old Chicago, Illinois, Lake Michigan, in Chicago Harbor. Established 1876.

Evanston, Illinois, Lake Michigan, on the grounds of Northwestern University. Established 1876.

Kenosha, Wisconsin, Lake Michigan, in the Kenosha Harbor on Washington Island. Established 1876.

Racine, Wisconsin, Lake Michigan, in the harbor. Established 1876.

Milwaukee, Wisconsin, Lake Michigan, near the south side of the harbor entrance. Established 1876.

Sheboygan, Wisconsin, Lake Michigan, at the side of the harbor entrance, Established 1876.

Two Rivers, Wisconsin, Lake Michigan, at the entrance of the harbor. Established 1876.

Kewaunee, Wisconsin, Lake Michigan, north side entrance of the harbor. Established 1894.

Sturgeon Bay, Wisconsin, Lake Michigan, north side of the east entrance to the canal. Established 1886.

A Brief History Of The United States Coast Guard

By the early nineteen hundreds, the federal government looked to unify its federal maritime agencies: the Revenue Cutter Service the Life-Saving service, the Steamboat Inspection Service, the Navigation Bureau, and the Lighthouse Establishment. They all dealt with nautical issues but were under the direction of different departments of the government. In 1915 all of the various maritime functions were organized into one department named the United States Coast Guard.

The Coast Guard has changed with the times. Their boats improved as technology advanced or the need arose. During World War I the Coast Guard's assets included fifteen cutters, and less than six thousand men. The cutters were used to escort convoys taking military supplies across the Atlantic Ocean. Their duties in the escort were to detect and eliminate underwater threats. The ships were equipped with depth charges used to destroy enemy submarines.

After the war, the Coast Guard was called on to enforce the nations Prohibition laws. During the early 1920s most of the Coast Guards efforts were along the Atlantic coast where rumrunners anchored beyond the three-mile territorial limit where the United States Coast Guard did not have any authority.

The bootleggers sailed to various Caribbean islands or north to the Canadian Maritime islands, filled their ships with beer, rum, whiskey or French wines and sailed back to the coast of the United States. The most lucrative markets were off New York, Long Island and New Jersey where the "Mother Ships" anchored, and smaller boats, called contact boats, raced out to buy the liquor. Then, usually under the cover of darkness, the smaller boats sailed back to the thirsty residents in the States. The idea

was nothing new, often ships from Europe anchored off shore and smaller craft took on loads of cargo and smuggled it to shore without paying the custom duties.

Arguably the most well known of the Atlantic rumrunners was Captain William McCoy. Captain McCoy developed a reputation of a gentleman bootlegger. He only sold quality liquor and at fair prices. The phrase, "The Real McCoy," meaning the genuine item, is said to stem from the fine booze from Captain McCoy.

The Coast Guard chased and intercepted some of the small boats bringing in the liquor but they were out numbered by those willing to take the chance and carry the booze to a market that was willing to pay top dollar.

To counter the ships off the three-mile territorial limit, the United States Congress passed a law moving the territorial limits of the States to twelve miles. The ships had to stay out further and the small boats had to make a much more dangerous trip to purchase the liquor. There were many small boats that did not return from a nighttime rendezvous at rum row but there were still enough bootleggers chasing the big money the occupation offered. An $8.00 case of whiskey was sold for $65.00 at rum row.

The main problem the Coast Guard had with intercepting the rumrunners was that the bootlegger's boats were much faster than those of the Coast Guard. The government cutters were built for endurance rather than speed and the other Coast Guard boats of the time were harbor tugs and surfboats that were not equipped to chase the speedy bootlegger's craft.

In 1923 the Coast Guard received the funding for 3,500 additional men and the construction of 38-foot picket boats capable of operating in shallow water and fast enough to run down the rumrunners. The other Coast Guard craft that were built for the war with bootleggers were the "Six-Bitters". They were seventy-five feet in length and built to operate off shore. The ship was used to harass the mother ships and to intercept boats carrying supplies to the mother ship and booze back into the States.

The vessels were built by American boat builders, many of which were at the same time building boats for individuals involved in the rum running business. When the Coast Guard intercepted a fast contact boat they confiscated it and added it to their fleet using it to chase down other illegal loads.

About the mid 1920s, the Coast Guard altered their main efforts in the war with the bootleggers from the Atlantic coast to the Great Lakes. The

The Curtis Flyingboat was the aircraft of choice when the Coast Guard entered the world of aviation. From the United States Coast Guard Historian's office.

Great Lakes were a rumrunners dream, close to Canadian distilleries and breweries and little or no law enforcement to block their way. Coast Guard vessels were transferred from the Atlantic coast to the Great Lakes and proved to be somewhat effective in the battle with bootleggers but the money that could be made in the trafficking of illegal booze kept the bootleggers coming. When one was captured there were usually two to take their place.

The Coast Guards involvement World War II was varied and active. The control of the Coast Guard was transferred from the Treasury Department to that of the Department of the Navy. Cutters were reconfigured with heavy armament and depth charges to escort convoys of ships across the Atlantic, along the US coast, the Caribbean, the Gulf of Mexico, North Africa and the Mediterranean.

The Coast Guard Cutter *U.S.S. Alexander Hamilton* was the first United States warship that was attacked and sunk after the invasion of Pearl Harbor. Twenty-six of her crew were killed in the attack.

After the close of World War II the Coast Guard went back to their main focus of ensuring the safety of sailors on the territorial waters of the oceans and Great Lakes. The assets they used have greatly improved as technology advanced. They went from wood surfboats powered by the muscle of men pulling at the oar and cutters powered by sail, to gasoline powered surfboats and huge destroyer type cutters capable of operating in most any sea conditions.

The Coast Guard became involved in aviation in the early 1900s. In 1915 two Coast Guard officers experimented with aircraft for use in coastal protection and search and rescue. Their work was so successful that by 1916 Congress authorized the Coast Guard to establish ten air stations. The first aircraft used were Curtis Flyingboats. The Flyingboats were used to fly over the surface of oceans and lakes searching for rumrunners making their way back from the mother ships or crossing the lakes from Canada. During World War II Coast Guard aircraft were used to locate survivors of shipwrecks. From its aerial position the aircraft could search a larger area than by boat and located thousands of survivors and directed support vessels to their location.

The Coast Guard took advantage of advances in aircraft development and was a forerunner in the use of the helicopter. An aircraft that can lift off and land vertically and hover over an object was of great interest to the Coast Guard.

After World War II there was a huge increase in recreational boating and the Coast Guard saw an increased need for their services. A helicopter could lift entire crews from a boat in trouble, deliver pumps for removing water from sinking vessels or fuel to those that had run out.

The Coast Guard began in 1790 as the Revenue Service under the Department of Treasury, in 1915 the Revenue Service merged with the Life-Saving Service then taking the name Coast Guard. In 1939 the lighthouse service became part of the Coast Guard. The Coast Guard was moved to the Department of Transportation in 1967 but in 2003 it was moved to the newly created Department of Homeland Security.

Today the United States Coast Guard is a multi-faceted organization charged with enforcing maritime law, mariner assistance, search and rescue, marine environmental pollution, lake, coastal and river aids to navigation, and it has extensive responsibilities under the Department of Homeland Security.

The Wreck Of The Henry Cort

The history of the Great Lakes is filled with many unusual events and unique watercraft. Arguably some of the most unusual ships to sail the inland seas were those designed by Captain Alexander McDougall. His designs remain known as the "whalebacks."

The whalebacks were characterized by their cigar shaped hull and cone shaped ends. The bows of the whalebacks were raised above the surface resembling a pig's snout and causing many to refer to the boats as pigboats.

The advantage of the design was that the ships could be built quickly and less expensively than their traditional counterparts. The design of the ship also allowed the ship to operate in heavy seas at a faster speed and their raised bow was excellent for operating in ice.

During the time period of 1889 until 1898, 43 whaleback-designed vessels were built. That included 19 steamers, 23 barges and one passenger ship, the *Christopher Columbus*.

The design was widely accepted but in time its deficiencies proved to be its downfall. The major design flaw that ended the production of the whaleback design was that it could not be off loaded by the new shore side unloading equipment.

Two of the whalebacks built at Captain Alexander McDougall's shipyard in Superior, Wisconsin in 1892 were the *Pillsbury* and the *Washburn*. The two were engaged in the package freight trade to ports around the Great Lakes. Three years later the two sister ships were sold to the Bessemer Steamship Company, part of the Rockefeller vast holdings. The *Pillsbury* was renamed the *Henry Cort* and the *Washburn*, the *James B. Neilson*. The ships sailed under their new company for six years until the Bessemer fleet was taken over by the Pittsburgh Steamship Company.

The *Henry Cort* transported iron ore from Lake Superior to the steel mills on the Lake Erie. For the return trip the *Cort* often hauled coal from Lake Erie north.

December of 1917 was unusually cold. Several days of temperatures hovering around zero caused the St. Marys River, the St. Clair River, the Detroit River and shallow Lake St. Clair to freeze over. Thirty three steamers became stranded in the ice between the southern end of the Detroit River up to the northern end of the St. Clair river alone. If the ships were not freed, they would be subjected to the forces of the compacting ice and severely damaged or totally destroyed. The cost to Great Lakes shipping would be catastrophic.

The government called out all available ships with ice breaking capabilities to aid in breaking the frozen ships free. Ships such as car ferries with iron plated hulls for operating late into the winter steamed to the areas of the worst congestion.

The *Henry Cort* was at Lorain, Ohio, being prepared for winter when the call came for the *Cort* to assist in the ice breaking duties. The design

Captain McDougall's whaleback passenger ship the Christopher Columbus. *From the Muskegon County Museum.*

The Henry Cort *prior to the front turrent being removed. From the Muskegon County Museum.*

of Captain McDougall's whalebacks made them excellent ice breakers. The severe rake of their bow allowed them to ride up on the ice and the weight of the ship would break the ice.

The *Cort* steamed from Lorain towards the ice blockage at Bar Shoal on Lake Erie near the mouth of the Detroit River. She drove her bow up high on the pack ice, broke it, then backed off and drove forward to climb up the ice again.

A string of ships bound down the Detroit River broke their way through the pack ice. The ships struggled with the ice flows but the convoy kept moving to keep the channel open while the *Cort* worked off channel to free a stuck steamer.

The *Cort* plowed up on the ice towards the steamer, the ice cracked under the weight of the whaleback, then Captain Murphy ordered the engines astern to back up into clear water. The captain called for all ahead forward and the racked bow of the ship climbed up on the ice once again. The *Cort* would back into the channel cleared for the passing convoy so Captain Murphy had to time his charge on the ice.

The process proceeded until the *Henry Cort* climbed high on a particularly hard section of ice and slid backward off the ice, right into the path of the passing 600-foot *Midvale*.

Captain Murphy screamed for the *Cort* to be put into all ahead forward while the captain of the *Midvale* called for all astern. Neither action was enough, the *Midvale* smashed into the *Cort*. Cold Lake Erie water poured into the *Cort*. Steam whistles blasted a distress signal as the *Midvale* backed away from the *Cort*.

A car ferry breaking ice nearby steamed towards the *Cort* and took on its crew as the whaleback slowly sank to the bottom. With the ice conditions there was no way to salvage the whaleback. It would have to wait till spring.

In April, a salvage crew went to the position of the *Henry Cort* but the ship wasn't there. The wind blown ice moved the 320-foot ship several miles from her last known position. Throughout the summer repairs were made, the ship was pumped out and raised.

The *Henry Cort* was once again a working ship on the Great Lakes. For the next seventeen years the whaleback went through some structural changes; the forward turret crew quarters was removed and rebuilt, the after house was rebuilt and a crane was installed. The crane was mounted on rails on the ship's deck to be used to remove cargo. When the cargo was scrap steel or pig iron, the crane was equipped with an electromagnet, or a clam bucket was used for grain or coal.

During 1933, the *Cort* met with more troubles. While making its way through the Muskegon Channel, the *Henry Cort* struck a small outboard motorboat. The three fishermen aboard the boat were cast into the water. Two of them drowned before help could arrive, the third fisherman was able to swim to a life ring thrown from the *Cort*. Federal Steamboat inspectors investigated the incident and cleared the captain and crew of the *Henry Cort* of any wrongdoing in the incident. Although in 1933, the *Cort* again met the bottom. While powering through the Detroit River she scraped the bottom and damaged her hull. A tug went to her assistance and escorted the *Cort* as she raced for a dock, but the old girl sank just short of the dock. She was once again raised and put back into service.

In 1934, the *Cort* now in her 42nd season had been relegated to hauling scrap steel and pig iron from smaller ports to foundries. The transportation of iron ore, coal and grain was handled by the larger more economical steamers, which could carry in one trip ten to fifteen times the cargo capacity of the *Henry Cort*.

In late November, the *Henry Cort* took on a load of 2,000 tons of pig iron in Chicago bound to the Holland Furnace Company in Holland, Michigan. After offloading its cargo, it was headed back to Chicago for more pig iron. But a storm was descending down on Lake Michigan.

Winds with gusts exceeding 60 miles-per-hour whipped the lake into monstrous waves.

The old whaleback plowed into the waves trying to maintain headway, but whalebacks were not known for their ability to power into the seas. The ship's unusual bow tended to be blown off course. Captain Charles E. Cox, master of the *Cort*, thought of coming about and running before the storm but feared that the ship might capsize if he came abreast to the seas.

The *Cort* continued south towards Chicago barely making headway in the heavy waves and gale force winds. When the ship was off Michigan City, Indiana, the wind and waves battered the old whaleback and pushed her into the trough then whipped her around until she was heading back north.

Now the waves were breaking down on her stern as she powered back up along the Michigan coast. Captain Cox knew his ship was in peril. If he

A photograph of the Muskegon Channel taken from shore looking out into Lake Michigan. From the Muskegon County Museum.

continued with the limited control he had over the *Cort*, he was sure they would end up grounded on a reef and beaten to pieces by the relentless waves. Or, he could try to make the harbor at Muskegon, Michigan.

There was no choice to be made, Muskegon was the only chance for the *Henry Cort*. Captain Cox ordered the old girl to starboard and made for the narrow Muskegon Channel. The ship took the storm on her starboard side as she steamed east toward the safety of the harbor.

The *Cort* held up well against the pounding she was taking from the storm. About 10:00 pm on November 30, 1934 as they neared the breakwalls on the north and south side of the Muskegon Channel the waves combined with the backwash off the breakwall, lifted the *Cort* and slammed her down on one of the large boulders of the north Breakwall.

The ship's keel was broken and many of her plates buckled and broke. Cold lake water poured into the engine room and cargo hold causing the lights to be extinguished and the ship to settle to the bottom. The waves that broke on the north breakwall smashed down on the wreck of the *Cort*, continually pounding the ship.

The Muskegon Coast Guard Station received a report of a ship on the breakwall and notified the Tenth District Headquarters in Grand Haven, Michigan, then readied their lifeboat to go to the aid of the ship. The Grand Haven Station surfboat and the Coast Guard Cutter *Escanaba* were made ready and put on standby.

The site of a ship smashing on the Muskegon Channel breakwall reminded many of the older population of the October, 1919 wreck of the Steamer *Muskegon*.

Debris lines the shore after the Muskegon *broke up on the Muskegon Channel north breakwall. From the Muskegon County Museum.*

The *Muskegon* was making its way into the channel during a storm and it too was lifted high by a wave and slammed down on the breakwall. The *Muskegon* was hung up on the south breakwall and the waves from the north beat the ship to pieces which were scattered along the channel and Muskegon Lake.

Now fifteen years later, the Muskegon Coast Guard was alerted that another freighter was breaking up, this time on the north channel breakwall. Chief Boatswains Mate, John Basch, in command of the Muskegon Station, and four of his surfmen, Charles Bontekoe, Edwin Beckman, Roger Stearman and Jack Dipert, set out from the station in their lifeboat powered by a 40-horse power gas engine. They braved the 60 mile-per-hour winds, the current of the water rushing through the channel from Lake Michigan, and the punishing waves and backwash from the breakwall.

The five men in their underpowered boat battled against the current to the end of the breakwall and rounded the north wall towards the ship. The waves washed over the small boat, threatening to overturn it and throw the men into the cold lake.

Chief John Basch, a veteran of 25 years service, later said; "The run into the waves driven before a 60-mile gale while we were attempting to get out the *Henry Cort* was one of the worst experiences in my life."

The small boat rounded the north Muskegon Channel breakwall seeking the calmer waters on the lee side of the ship. Calmer being a

EXTRA THE MUSKEGON CHRONICLE EXTRA

77TH YEAR — MUSKEGON, MICHIGAN, SATURDAY, DECEMBER 1, 1934 — EIGHTEEN PAGES — THREE CENTS

FREIGHTER CORT DOWN; CREW IS BELIEVED DEAD

COAST GUARD MEMBER LOST OVERBOARD

WASHED FROM BOAT GOING TO SINKING CORT

CAPT. PRESTON ARRIVES WITH CREW OF SIX

RESCUE TRY IS MARKED BY HEROISM

LIST OF CREW ABOARD CORT

SINKS AFTER HITTING BREAKWATER; FATE OF CREW NOT DETERMINED

relative term, the lake was still quite rough with the waves breaking over the wall and the wind blowing at near hurricane force velocity.

The Guardsmen searched the freighter being pounded on the rip rap (large rocks placed along the breakwall) for signs of life. They saw no movement or lights aboard the ship.

Chief Basch was at the wheel, strapped into a harness that held him in place so his hands were free to pilot the boat. The boat rose and fell on the waves and violently rolled from port to starboard. The faces of the men were reddened and stung from the wind blown spray but they knew the task at hand was to save the crew of the ship before it completely broke up. As Chief Basch slowly made way towards the *Cort*, one of his men screamed out, "Man overboard!"

Jack Dipert, a 23-year-old member of the Muskegon Coast Guard Station, was washed out of the boat by a wave breaking over the boat. Jack Dipert was in his first year with the Coast Guard but he knew the service well. His father, William Dipert, was a twenty-seven year veteran of the Life-Saving Service and the Coast Guard and officer-in-charge of the Point Betsy Station.

The men sent to rescue the crew of the freighter now were searching for one of their own. Chief Basch yelled over the roar of the wind and waves to keep a sharp eye out as he began to come about in the monstrous seas. The spray blown by the sixty mile-per-hour gale and the dark of night hampered their search but each man called out for Jack and prayed he would soon be seen floating on the crest of the next wave. Chief Basch recounted that they saw Jack Dipert after he was washed out the boat for just seconds, then he disappeared in the darkness and huge waves.

The men of the Muskegon Coast Guard Station braved the wind and tremendous waves searching for their missing comrade until a wave broke over their boat and swamped their engine. The lifeboat was now at the mercy of the wind and seas.

The men drifted north of the breakwall towards shore until their boat grounded and the waves flipped it over and the men were thrown into the raging waves. They were tumbled by the waves towards shore held afloat by their cork jackets.

Once on the beach, the men dripping wet and exhausted from the ordeal began walking along the beach searching for Jack Dipert until they were picked up and taken back to the station where they changed into dry clothes then went to the beach near the north breakwall.

At about 11:00 p.m., the Tenth District was informed of the missing Guardsman and the failed attempt to rescue the crew of the *Cort*. Lieutenant Ward Bennett, Commander of the Tenth District, ordered the cutter *Escanaba* and Captain Preston of the Grand Haven Station and five of his men to go to Muskegon, 12 miles up Lake Michigan.

Captain Preston and his men ventured out into the open water of Lake Michigan in their lifeboat powered by a 90-horse-power gasoline engine. They battled huge waves as they made their way to Muskegon. Many times the men thought their small lifeboat would capsize under the pounding it was taking. Several seams of the boat parted and it took on water but the pumps were able to keep up. The men later related that it was the worst conditions any of them had ever been in.

Once they reached the Muskegon Station, they took on dry clothing and met with Chief Basch to determine the next course of action. The *Escanaba* arrived and moved as close to the *Henry Cort* as they safely could. There was no sign of life aboard and most thought the 25 man crew had perished in the accident. The Coast Guard cutter shined its powerful spotlight on the darkened hulk of the *Cort* beating against the breakwall with each heaving sea. It looked deserted. Then a small light shined back through a portlight. Someone was alive. There was at least one soul to be saved.

With the seas breaking over the wall and pounding down on the ship, the *Escanaba* could not get close enough to the *Cort* to affect a rescue. It reported to the Muskegon Station seeing the light on board then the cutter went off shore to wait for daylight and to ride out the storm in deeper water.

The day after the whaleback Henry Cort *was pounded on the North Muskegon breakwall, shown, the battered and broken ship sunk to the bottom. From the Muskegon County Museum.*

With the news that at least some of the crew of the *Cort* were still alive, Chief Basch and his men set out on another rescue attempt. Dressed in their slickers, they began walking along the north breakwall towards the wreck. This was not an easy task considering the huge waves breaking over the wall. The men had to time the waves and hurry forward between waves, then hide behind a rock as the wave crashed over them. If they would be caught in the open when the wave hit they would have been washed off the wall into the lake. One man struggled as he carried a coil of rope over his shoulder along the length of the breakwall.

Slowly they made their way until they were abreast of the heaving ship, rising and smashing down on the rocks of the breakwall. Once as near to the ship as they could get, they shot a light line attached to a heaver line to the *Cort*. A crewman on the *Cort* braved the waves that crashed down on the ship and pulled on the light line until the end of the heavy line was aboard the ship. The heavy line was secured to the ship and to the breakwall.

The only way for the crew to get off the *Henry Cort* before it broke to pieces and threw the crew into the lake to a cold death was to hang from the line and move hand over hand from the ship to the rocks of the breakwall. As dawn began to provide light, the first man gripped the rope tightly and with great fear left the rocking ship. He let go with one hand

and quickly placed it in front of the other, then the other hand until he was close enough for the men of the Muskegon Coast Guard to reach out and pull him onto the breakwall.

One by one all of the 25-member crew of the *Henry Cort* made their way off the wreck of their ship to the north Muskegon Channel breakwall. The thousands of people who had gathered on shore to watch the dramatic rescue cheered as the last person, Captain Charles Cox, was lowered to the breakwall.

They were off the ship that was breaking into pieces but they were not yet safe. They still had to walk the 1/10th of a mile of the breakwall while waves violently smashed into it and a torrent of cold lake water cascaded across the wall.

The Guardsmen lashed themselves and the crew together with a rope and set off along the breakwall avoiding the crashing waves and hurrying across the breakwall through the retreating water. Some of the crew were injured and were carried by others. The people on shore shouted encouragement as they watched the men slowly make their way along the breakwall.

Once the parade of cold and injured men were close to shore, bystanders rushed out on the breakwall, daring the waves, to help the men to shore.

The wreck of the Henry Cort *from the bow as she lays wrecked on the north Muskegon Channel breakwall. From the Muskegon County Museum.*

THE WEATHER

THE MUSKEGON CHRONICLE

FINAL EDITION

77TH YEAR — MUSKEGON CHRONICLE, SATURDAY, DECEMBER 1, 1934 — FOURTEEN PAGES — THREE CENTS

SEEK BODY OF COAST GUARDSMAN AS CREW OF 25 ABOARD CORT RESCUED

C. M. GREENWAY, HEAD OF BOOTH PAPERS, EXPIRES

President of Michigan Group of Dailies for Past Two Years

CAREER REMARKABLE

Started as Cashier for Old Grand Rapids Press in 1893

DIES

AID TO COUNTY WILL CONTINUE

Recommendation to Be Made

PRESIDENT ON THRESHOLD OF RELIEF RULING

Decision to Be One of Most Momentous Ever to Face Chief Executive

CITIZENRY IS DIVIDED

Whether to Taper Off Federal Expenditures Is Question

VICTIM OF STORM

SAILORS REST IN CCC CAMP AFTER BEING SAVED FROM VESSEL

Hundreds Throng Beach Near Scene of Wreck on Breakwater of Whaleback Freighter; Two of Crew in Hospitals from Exhaustion.

BLAMES LIGHT LOAD IN WRECK

Two of the crew were taken to area hospitals with broken bones, exhaustion and hypothermia. The rest were taken to the Muskegon Civilian Conservation Corps (CCC) Camp where they received dry clothes and hot meals.

The wreck of the *Henry Cort* brought out the bravery of many: Chief Basch and the men of the Muskegon Coast Guard Station who heroically set off in horrendous conditions to save the lives of the crew of the *Cort* and then ignored the risk of their own lives as they walked the breakwall to the ship. The men of the Grand Haven Coast Guard Station who made the trip from their homeport to Muskegon through one of the worst storms in memory. The captain and crew of the Coast Guard Cutter *Escanaba* that unquestioningly left their dock when most vessels were heading for the safety of harbors. The men of the CCC camp who walked the beach during the terrific storm searching for the body of Jack Dipert while the beach sand blown by the 60 mile-per-hour wind sand blasted their faces. The brave individuals who left the safety of the shore to venture out onto the breakwall to assist in bringing in the survivors off the wreck of the *Henry Cort*. Finally Jack Dipert, the 23 year-old Coast Guardsman, who bravely left the Muskegon Station knowing he might not return. He paid the ultimate price in a heroic attempt to save the lives of others.

Billy Gow: Hero Of The Tug *Reliance*

To save the crew and passengers aboard the tug *Reliance,* it took the daring of many, but if it were not for the skill, fortitude, strength, and ingenuity of crewman Billy Gow all would have been lost.

Twenty lumbermen, who had been working in the northern Ontario forests on the north side of Lake Superior, waited to board the tug *Reliance*. Standing at the dock in the small lumbering town of Puckasaw, Ontario, the men were dressed in heavy woolen coats, their chooks and wool caps pulled down over their ears against the brisk winds on the Lake Superior shore.

They worked in the northern Ontario woods but it was colder near the shore, damper and much windier, than in the deep woods where they labored. The men didn't seem to mind and their spirits were high because they were about to board the ship that would take them home to their families for the Christmas holiday.

The tug *Reliance* was considered a seasoned vessel with thirty years on the lakes; the wood tug had been through a lot. Built in Collingwood, Ontario, in 1892 at the Collingwood Dry Dock, the tug was 124-feet in length and 23-feet in beam.

On that December morning of 1922, the tug carried her normal crew of 14, the 20 lumbermen and two other first class passengers.

One was Captain John McPherson, a prominent man of Sault Saint Marie, (pronounced Soo Saint Marie) Ontario, who worked for the Booth Fisheries. The captain had traveled to the company camps making sure that winter supplies were delivered and that the men at the camps had all that they needed to make it through the winter. Captain McPherson had boarded the *Reliance* to return home.

The last of the passengers was Fred Regen, who worked in the Forestry Department of the Lake Superior Paper Company. He was retuning home after supervising operations in the Puckasaw River area.

The tug Reliance *in the Soo Locks with a log raft. Photograph courtesy of the Le Sault de Sainte Marie Historical Sites, Inc.*

That made 36 passengers and crew aboard the *Reliance*, all anxious to celebrate the Christmas holiday in their own homes with their loved ones.

The course the ship would follow would take them along the north shore of Lake Superior between the mainland and Michipicoten Island. Then she would change to an east southeast course into Whitefish Bay, then into the St. Marys River and on to Sault Saint Marie.

Ice was already forming along shore and in the rivers and it wouldn't be long before the big lake was frozen for miles from the shore, so the tug's master, Captain D. A. Williams was eager to get this trip over.

As the wood tug *Reliance* steamed away from the wharf at Puckasaw, the weather was cold with snow flurries and a moderate wind, but nothing to indicate the severe weather they would soon encounter.

While in route it began snowing harder and the wind increased until it was blowing a full gale. The *Reliance* plowed through a blinding snow storm of blizzard proportions and battled huge storm waves for most of the 50 miles the tug had covered since departing Puckasaw. The helmsman, unable to see much more than a boat's length ahead, was relieved when Captain Williams had had enough of the storm and decided to seek shelter at Gargantau Harbor.

The harbor was a lumbering town and a harbor of refuge for ships traversing the Lake Superior north shore, but now this late in the season, most residents had left the isolated area. The stores were closed and boarded up and there wasn't any fuel to be had. As many towns in the north are during the dead of winter, Gargantau was a virtual ghost town.

The passengers and crew remained on the tug *Reliance* huddling around the stove in the galley and the firebox in the engine room to keep warm and settled in for the night. The storm blew all night and well into the next day.

The winter storm assaulted upper Lake Superior with steady gale force winds, snow was measured by the foot and seas reaching monstrous proportions continued to rage on the lake, forcing the *Reliance* to remain in the harbor for four very cold days.

On Wednesday, December 13th chief engineer Walsh reported to Captain Williams that if they stayed in the harbor much longer they wouldn't have enough fuel to make it to Sault Saint Marie. They were burning a large amount of fuel to keep the stoves hot to keep themselves from freezing.

The captain inquired of Jack Hartens, who, along with his wife were the ship's cooks, if the food stocks were a concern. He was told that the supply of food was low. The extra days in route and the large amount of passengers were taxing the stores.

Captain Williams realized that they must leave the safety of the harbor and set out for the Sault before it was too late and they became stranded in Gargantau Harbor.

In the early morning hours of Wednesday, December 16, 1922, Captain Williams found the wind and snow had not stopped but it had subsided and decided that it was time to leave.

The captain and mate talked over their options and determined they could skirt along the Lake Superior shore, just far enough out to avoid the shoal water. That way they would avoid the heavy seas in open water and if the storm were to blow again they could quickly run to the safety of a cove or the lee of an island.

The crew was busy performing their duties, firemen shoveled coal into the firebox, and oilers kept the engine well lubricated, the Hartens prepared lunch for the 36 aboard, the wheelsman held the wheel tightly, fighting the urge of the boat to be blown off course by the waves and wind from the southwest, while the captain kept a vigilant watch out the pilothouse windows.

Throughout the morning the *Reliance* steamed east but they had not been out long before the storm again grew into a gale. The wind blew and gusted with hurricane velocity. The captain stayed in the pilothouse quietly questioning whether the tug was still on course; with the snow obscuring his visibility he wasn't able to take a bearing. The compass heading seemed right but he would have been reassured if he could see land to ascertain how far from shore they were. The area they were sailing through was scattered with islands and reefs but the course they set should put them out far enough from shore not to worry about them.

The wind blew the snow horizontally across the ship and the waves broke hard on the starboard of the *Reliance*. Not a day to be out on Lake Superior.

While the passengers huddled around the stoves trying to stay warm and listening to the howling wind, the tug *Reliance* suddenly ground to an abrupt stop as she plowed hard onto a shallow reef.

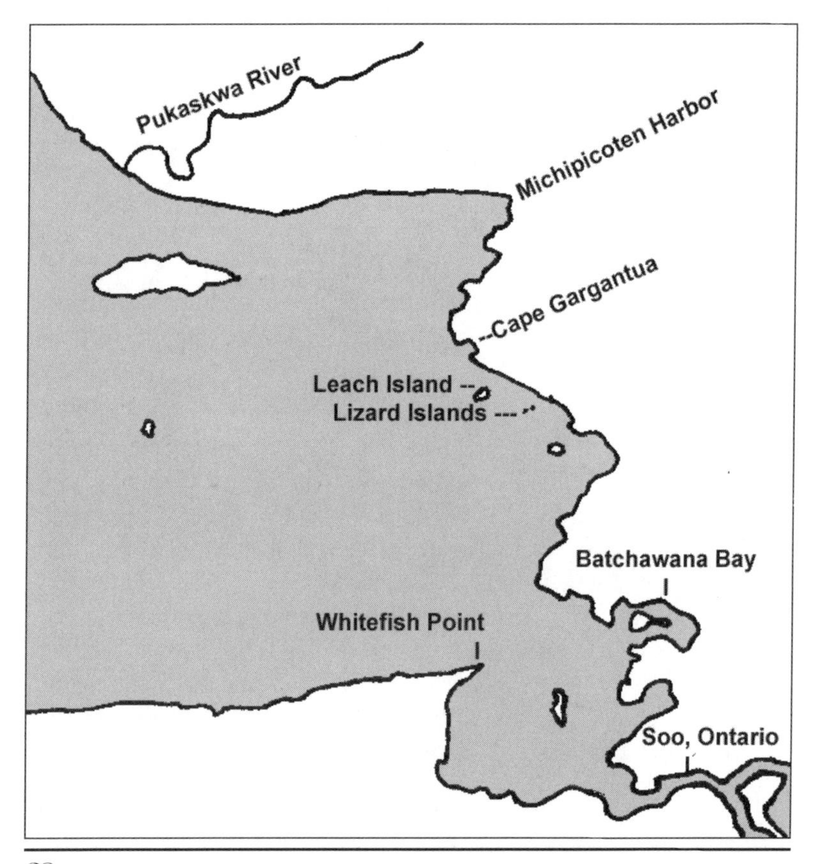

A photograph of the tug Reliance *grounded on Lizard Island. From the Edwin Brown Collection of the State of Michigan Archives, Lansing, Michigan.*

The sudden stop sent passengers and crew falling. Some fell to the floor while others toppled on top of them. In the engine room the men fell to the floor, instinctively reaching out for something to break their fall. In the galley pots, pans and dishes tumbled out of their racks crashing to the floor. Anything not secured flew about crashing into walls, the floor and the scattered bodies.

The tug had run up on the shallow reefs surrounding the Lizard Islands; a group of small uninhabited islands about 4 miles off the Canadian mainland.

The *Reliance* was swung broadside in the waves. Momentarily riding the crest of a wave, the tug was thrown hard against a large boulder smashing a hole in her port side.

Huge waves blown by the severe gale winds crashed down on the *Reliance*. Captain Williams rang for all astern full but the chief engineer yelled that the propeller was gone and the shaft had bent when it smashed on the rocks.

Fireman Gow yelled above the roar of the storm to the chief engineer that the hull had been holed by the rocks and the ship was taking on water. The tug *Reliance* was not going anywhere.

Everyone on board knew they were in trouble. They couldn't stay onboard the wave drenched tug and leaving the ship in a raging blizzard was suicidal, but leaving the sinking tug was their only option if they wanted to stay alive. It was just a matter of time before the waves beat the tug to splinters.

Aboard the *Reliance* were two lifeboats and a yawl. On the aft where the yawl was kept passenger and crew readied the boat to be lowered. It was on the portside of the tug in the lee of the wind and somewhat protected from the wind and waves. Eight people frantically climbed into the yawl while the remaining passenger and crew readied the two lifeboats located forward.

The lifeboats were larger than the yawl; each had a capacity of carrying 18 persons.

Captain McPherson went to the yawl and tossed in some blankets he had gathered below. He also tossed down his grip into the yawl, the boat was already filled and the waves were forcing it away from the *Reliance* so Captain McPherson shouted over the roar of the storm that he would go with one of the other boats.

The constant pounding of the waves on the starboard side of the tug severely rolled the vessel from side to side. One large wave crashed down on the side of the ship, another huge wave following in close succession smashed into the tug rolling the *Reliance* over on her port side. Icy cold Lake Superior water poured into the cabin.

Men on the ship leaped to starboard, grabbing onto anything to keep themselves from being thrown into the water. Fortunately the tug remained on her port side in the shoal water rather than rolling completely bottom up.

The yawl, still tied to the tug with her eight passengers, almost overturned as water began to pour into the boat but luckily it remained upright. But the bow of the yawl was pinned under the *Reliance* and unable to get away from the rolling ship.

Assistant engineer Curry jumped into the bow of the yawl and pushed up on the overturned cabin forcing the yawl from under the tug. The yawl floated a few feet away from the *Reliance* when a wave picked it up and swept it away.

The nine now aboard the yawl looked back through the blowing snow at the overturned wreck of the *Reliance*. They could see crew scurrying to the lifeboats. Deckhand Billy Gow, without thought of his own safety, repeatedly went below in the overturned tug to carry provisions up from the galley they would need to live through this terrible ordeal.

Captain Williams had ordered Norman Stotes to send a distress message when they first struck the reef. The wireless operator was thankful to hear a reply indicating their message had been received, but

The tugboat General. *Photograph courtesy of the LeSault de Sainte Marie Historical Sites, Inc.*

the radio failed moments later. The water rising in the ship flooded the dynamo room preventing it from generating the electricity the wireless needed. The radio was dead.

At Sault Saint Marie, the emergency message was received and the tugs *Gray* and *General,* realizing they were the only vessels close enough to reach the tug before the crew and passengers froze to death, set out to rescue the men of the grounded tug.

The yawl was wildly blown by the wind and waves. Its passengers hoped they were headed towards shore but the blizzard conditions obscured any sight of land. They were buffeted by the snow, wind and waves for almost an hour before the yawl grounded to a stop on the mainland.

The passengers climbed out of the boat into the icy waters and trudged through the water and ice to the beach covered with deeply drifted snow.

The nine from the yawl dug through the waist deep snow for driftwood and sticks to start a fire; a fire for warmth but also to guide the other two lifeboats to shore.

Mrs. Hartens placed her wet shoes near the fire, unfortunately too close; they caught on fire and were burned beyond use. She would have to walk through the snow without shoes.

The nine survivors waited by the fire for the two other lifeboats to arrive on shore. The men took turns leaving the warmth of the fire to walk the beach into the heavy blowing snow looking for their comrades in the lifeboats. They looked but the lifeboats never came.

"Maybe they were blown further down the beach?"

"Maybe they couldn't get the boats lowered."

Everyone had a theory why the other boats were late in beaching, but none of them wanted to say what was on everybody's mind: the lifeboats might have capsized and their friends cast into the icy cold water of Lake Superior to die a cold and painful death.

The emergency call from the *Reliance* was received on Wednesday but the southwest gale blowing delayed the *General* and *Gray* from leaving Sault Saint Marie.

When the storm slowly subsided, the tug *General* was able to steam out of the St. Marys River toward the Lizard Islands to rescue the stranded men. En route the tug encountered several down-bound ships which had become trapped in the fast forming ice. The *General*, a large tug capable of breaking ice and clearing a path, was called to duty, for if the ice grew thicker, the ships would be trapped in ice for the winter, dooming the ship to being crushed in the spring breakup by shifting ice flows.

The tug *Gray* was readied and left the Sault the following day. The tug traveled about 30 miles before catching up with the *General* near Batchawana Point. The crews of both tugs suffered from the and biting wind and cold, but they knew the lives of the passengers and crew of the *Reliance* depended on them.

A wireless from the *General* reported that she was short on fuel from the ice breaking and needed to turn back to the St. Marys River, but the *Gray*, replied they would go on.

The stiff gale from the southwest continued to blow. The captain and crew of the *Gray* were intent on getting to the Lizard Islands as soon as they could to rescue the men on the wreck of the *Reliance*, but the storm was too much for the *Gray*; she had to turn back and shelter behind Batchawana Point.

Onboard the *Reliance,* Captain McPherson, Fred Regen and others attempted to lower the starboard lifeboat. The lifeboat swung violently on the davits crashing against the hull. The line separated, and the lifeboat was picked up by a gust and thrown against Mr. Regen and Captain McPherson knocking the men from the tug into the cold water of Lake Superior.

The men aboard the *Reliance* watched in horror as the captain was struck in the back by the lifeboat and pushed into the lake. They threw a life ring but the captain's lifeless body made no attempt to swim towards it. The lifeboat had broken Captain McPherson's back.

Fred Regen swam towards the tug, frantically fighting the waves as he reached the tug, momentarily held on to it, but the cold of the water had taken his energy and Fred Regen's bitterly cold fingers loosened their grip. He simply slid below the surface of the churning lake.

Hopes dimmed but there was still one more lifeboat.

As the lifeboat was lowered, it was blown by the wind and beaten by the continuous onslaught of huge waves. Men numb from the cold clung onto the wildly bucking lifeboat until the line was pulled from the cold hands of the men on the *Reliance*. Their last chance to get off the stranded tug was blown away, empty.

Twenty three men were now trapped on an up turned tugboat, huddling against the wind and waves on a small portion of the starboard cabin wall. The ship pivoted on the port rail grinding on the rocky bottom as the monstrous waves continued to beat on the wrecked ship.

Onshore the survivors of the yawl had waited by the fire for several hours and finally admitted to themselves that the others were not coming. Without any food or firearms to kill game for food and the blizzard not showing any signs of letting up, the nine had to find safety or they would surely die a slow cold death.

The group gathered their strength and set off through the storm and deep snow, sometimes trudging through waist deep drifts towards a lumber camp they knew was down the shore.

They walked for hours through the deep snow and blizzard for almost 12 miles, when Mrs. Hartens, walking without shoes, only the socks and rubbers found in Captain McPherson's grip, finally succumbed to the cold and exhaustion and fell unconscious.

The men had been taking turns helping her, sometimes carrying her, but now she lapsed into a state near death. Mr. Hartens told the others to go on and send a search party back for them if they found anyone.

They argued but they all knew it was best for them to leave Mrs. Hartens. They would be able move faster.

Mr. Hartens sat down and leaned against a tree with pine boughs blocking the wind. He wrapped his wife up with him in his coat, held her tight hoping his body heat would sustain her, and settled down to await their return. He was determined to remain with his wife and share his love with her till the end.

The seven men continued through the snow, their pants, shoes and socks soaked from the snow melted by their body heat. They slowly made their way through the deep snow drifts for another four hours until they reached a camp occupied by the Bussineau family at Agawa, Ontario.

While they were warmed by the fire in the Bussineau cabin they told their horrific story of the boat wreck, their harrowing ride on the wind blown yawl and about the 18-mile walk through a snow storm. They told Bussineau about Mr. and Mrs. Hartens left in the forest and implored him to help them.

Bussineau and his son, without a thought of their safety, pulled on their Mackinaw coats and went into the woods to find the two ship's cooks.

Walking in snowshoes to support their weight on top of the snow, the two made their way through the blowing storm towards the location where the couple was left. Bussineau and his son yelled out to the Hartens and listened for a reply.

Through the howling wind the men heard Mr. Hartens faintly calling back. Bussineau and his son found the two still alive but frightfully cold, Mrs. Hartens near death.

The men carried, and at times dragged, Mrs. Hartens over and through the drifted snow back to their camp. Mr. Hartens was given hot coffee and dry clothes and soon, in the warmth of the cabin, Mrs. Hartens slowly regained consciousness but she was still in serious condition.

Onboard the *Reliance* the 23 men hung onto the ship on that little part of the tug which was not yet submerged. They suffered tremendous hardships from exposure to the storm and sea.

One lumberman, Walter Longpre, unable to find a dry spot to sleep was forced to lay in an exposed position. He awoke the next morning with a swollen, frostbitten foot.

Captain Williams knew they had to find a way off the ship or they all would die from the cold, starvation or from being washed into the icy water when the ship broke up from the pounding it was taking from the waves.

While the others slept, Captain Williams and his helmsman devised a plan to build a raft and float to shore. They took a cabin door off its hinges and lashed it to two oil barrels, making a very crude raft, but a raft they were willing to stake their lives on.

In the morning, the men on the wrecked *Reliance* watched as the helmsman and captain pushed the raft off the tug and jumped onto the wildly pitching and rolling raft.

The two men lay across the cabin door, now the deck of their raft, and held on. The waves and wind pushed the two men through the stormy morning towards Lizard Island a little more than one hundred yards distant.

Captain Williams and helmsman Fred Langhaud crawled from the lake to the island trying to get away from the crashing waves. They had hoped to find the lifeboat that drifted away but it had not washed up on the island. They wondered how could they get the rest of the men off the *Reliance* before she was rendered kindling by the waves.

The two men couldn't ride the raft back against the waves to the tug. They were trapped on the island. The storm showed no signs of letting up so they would have to spend the night on the island. They dug through the waist deep snow to find sticks and built a small fire. They didn't have food or shelter for protection from the storm and they suffered a terrible night.

In the morning, Billy Gow stripped timbers from the tug and fashioned another crude raft. He affixed a line to the craft and had three men get aboard. The raft and its passengers were let loose and carried to the island.

Onboard the *Reliance* the men pulled on the line bringing the raft back against the force of the wave and wind for three more men to board. This continued until everybody was on the island except Billy Gow.

Billy, a hero of the incident, continued to think of the group and what they needed to survive in such brutal conditions. He gathered all the provisions he could and lashed them to the raft, then played out the line allowing the waves to carry the supplies to the island.

As Billy pulled the raft back to the *Reliance* for his trip to the island the line broke, the waves caught the raft and washed it away, The crude raft floated towards shore until it was smashed into splinters on a rocky reef.

Billy was not a man easily defeated; he calmly ripped up boards from the ship's stairway and made another small raft. He lay along the piece of wood and practically surfed to the island.

After a terrifying four days on the grounded tug, the 25 men were now off the ship which violently rolled with every wave that smashed into her, but they weren't safe yet. Now they were on an uninhabited island some four miles from the mainland with few supplies, in a raging Lake Superior snow storm.

At Agawa, the nine survivors regained their strength and were warmed and nourished at the Bussineau camp. They sent word to the Sault that nine were all of the 36 people aboard the tug *Reliance* that lived through the harrowing experience, news which caused much concern for the loved ones of those not rescued.

Since Mrs. Hartens was still in such serious condition she and her husband stayed at the Bussineau camp while the other seven set out on foot to Frater, Ontario.

They walked the five miles to Frater where the railroad made a stop. The seven boarded the train and traveled to Sault St. Marie, Ontario, where they were recuperating at the Albion House.

The twenty three men trapped on Lizard Island shared the one axe and cut down trees and piled the wood on the fire to keep warm. They knew that the wireless message Norman Stote sent had been received and anxiously waited and prayed for someone to rescue them.

The first day on the island their spirits were up, since they had successfully gotten off the *Reliance*. But on the second day, Charles Salo, a lumberman, died, and the remaining men began to grow depressed.

The other lumbermen said Salo was ill when he boarded the *Reliance* and the trauma of the wreck, the wet raft ride to the island and the 18 degree below zero temperatures were just too much for him.

By the fourth day on the island the men were very depressed. They tried to hunt but the snow was so deep they couldn't find any game. They thought of loved ones waiting for them at home and longed to be with them for the Christmas holiday. The storm had eased up but the wind still blew cold and the men huddled close to the fire to stay warm.

On the morning of their fifth day of being trapped on the snow covered Lizard Island the men heard a ship's whistle blow. At first they didn't believe their ears. But they jumped with joy when they saw the tug *Gray* in the distance.

Seeing life on the island, boats were lowered from the *Gray*. The sailors cautiously rowed towards the island, avoiding the rocks and negotiating the breaking surf. The twenty two remaining men from the *Reliance* ran through the deep snow and into the cold Lake Superior water to meet the boats.

The men, many suffering from exhaustion and frost bite, climbed into the lifeboats, thanking God and the crew of the *Gray*.

After being fed restoratives, warmed with hot coffee and the hot stove, the rescued passengers and crew of the *Reliance* told of their terrifying ordeal and praised of the heroism and ingenuity of Billy Gow.

They told of how Billy fashioned a raft to get them to shore and had to make another to save himself. The bravery, courage and ingenuity of Billy Gow would never be forgotten by the passengers and crew of the *Reliance*. He was indeed a hero.

Heroes Of The Armistice Day Storm Of 1940

Storms have ravaged the Great Lakes since the retreating glaciers formed them. Some storms were fast moving and wreaked havoc on the region quickly and moved on, while others like the notorious storm of November 1913 stayed over the Great Lakes region for days. One storm which may not be the worst on record but would certainly rate in the top five was the storm of November 12, 1940… the Armistice Day Storm.

The storm system developed over Kansas and passed to the east to Minnesota and Wisconsin. The states were unexpectedly struck by severely dropping temperatures, sometimes a variance of sixty degrees, winds of 50 to 80 miles-per-hour and up to 26 inches of snow blown into drifts up to twenty feet high.

In Minnesota people were lured outdoors by unseasonably warm temperatures for November. By mid-day temperatures reaching in the sixty degree range were not unusual. Many people went to the lakes and swamps for a day of duck hunting. The storm started with some light rain which changed to sleet then to snow as the temperature began to drop. Suddenly the winds began to blow and increase in velocity until it was blowing a gale. The snow intensified to a blizzard. The duck hunters were trapped out in the fields and swamps unable to find shelter. Forty nine people were killed in Minnesota by the Armistice Day storm, most of them duck hunters.

Onshore the great winds ripped trees up by their roots, power and telephone poles snapped, factories and farm buildings collapsed. Many people were killed when buildings were blown down or electrocuted by fallen power lines. Roofs of buildings were ripped off leaving the building open to several feet of snow. The destruction onshore isolated towns, communications cut off, electricity terminated by downed lines and roads blocked by fallen trees and huge snow drifts

The storm packing hurricane force winds laid siege to Lake Michigan for twelve hours catching boats and ships out on the lake by surprise. All along Lake Michigan from its southern tip at Gary, Indiana to the Upper Peninsula of Michigan at the north, ships in open water were in jeopardy as the skies darkened as the storm raged in across the lake. Before the gale force winds the lake built into great waves from the west.

As the storm raged unobstructed across the open water of Lake Michigan, huge waves assaulted the ships traveling along the Michigan shore line. At Ludington, Michigan, the large car ferry *City of Flint* was attempting to make it inside the breakwater to the safety of the harbor when the wind and waves rendered the big ship uncontrollable and she was forced onto the beach about three hundred yards from shore.

The ship lying abreast to the shore rolled violently with the onslaught of the waves. Fearing the ship would be pushed by the waves higher up the beach and be beat to pieces, Captain Jens Vevang ordered the ship to be scuttled. As lake water poured in, the large car ferry settled to the bottom.

The men of the Ludington Coast Guard Station raced to the beach opposite the *City of Flint* with their Beach Apparatus and prepared the Lyle gun to be fired.

The Coast Guard practiced the procedure often, for when it became necessary to use it to save lives, it would be second nature. The Lyle gun fired a projectile attached to a line up to seven hundred yards.

The line fired across the deck of the *City of Flint* would be securely attached to the ship. Then a life ring set in a horizontal position with rubber pants was sent along the line on a pulley. A sailor on the ship climbed into the rubber pants and was pulled across the raging sea and bone crushing breakers to the safety of the beach to deliver the message that the ship was holding up to the pounding and that the crew and passengers numbering 47 would remain aboard the ship until the storm ceased and it was safer for all to leave.

In upper Lake Michigan near the Lansing shoals, two large freighters reported their positions and their conditions. The *Frank Billings*, 444-feet in length, had its pilothouse blown away, the radio room flooded and the helmsman injured. The other freighter was the 415-foot *Conneaut* of the Wyandotte Transportation Company. She reported her rudder and propeller were damaged and both of her anchors were dragging.

The tanker *Justin C. Allen* had a broken rudder cable and was drifting uncontrollably in the high seas. She displayed distress signals hoping someone could come to their assistance.

The 420-foot steel freighter *William B. Davock* had cleared the Straights of Mackinaw with a cargo of coal bound for a steel mill on lower Lake Michigan when she sailed into the brunt of the storm. There was no distress call from the ship; she simply disappeared in the night. The fate of the *Davock* was not known until bodies of sailors in life vests with the stenciled name *Davock* began washing up on the beach near Ludington.

Other bodies of sailors from the 380-foot Sarnia Steamship Ltd. ship, *Anna C. Minch* also began to drift ashore near Ludington giving rise to the theory that possibly during the storm the *Minch* and the *Davock* had lost visibility in the blizzard and collided and both ships went to the bottom. But most knowledgeable sailors agreed that both ships were probably overcome by the winds and huge accompanying waves.

(Authors note: The wreckage of the *Davock* was discovered off Pentwater, Michigan, in 1982 and it did not show any signs of a collision.)

None of the crew of twenty four sailors aboard the *Anna C. Minch* survived the storm nor did any of the thirty three crewmen of the steamer *William B. Davock.*

At South Haven, Michigan two fishing tugs, the *Indian* and the *Richard H.* were reported over due. Four Coast Guardsmen from the South Haven Station ventured out into the lake wracked in turmoil in search of the two boats. The fishing tugs were never seen again and the Coast Guard boat was feared lost as well. It wasn't until the storm subsided that word was sent to South Haven that the Coast Guardsmen were safe in South Chicago. They had searched for signs of the fishing tugs but found they could not return to port with the storm raging so they continued on to South Chicago.

The brave men of the South Haven Coast Guard Station, who risked their own lives to try to save the lives of others, were safe but the eight men aboard the two fishing tugs were not seen again until their bodies washed up near Holland, Michigan. Wreckage of the two tugs washed ashore on a Grand Haven beach.

The Socony Vacuum Company oil tanker, *New Haven* had departed East Chicago in route to Muskegon with twenty three crew and had not been heard from. The Grand Haven Coast Guard found an oar from the tanker and the Muskegon Coast Guard, while patrolling the beaches, found a life ring, a life raft and other flotsam marked with the ship's name. It was feared that tanker *New Haven* had succumbed to the storm. Forty eight hours later the battered tanker arrived at East Chicago. One of the crewmen aboard the ship summed up their ordeal in four words; "We've been through hell!"

At the Little Point Sable Lighthouse, Keeper William Kruwell observed a freighter drifting in the storm tossed waves. Coast Guardsmen from Pentwater, Muskegon and Grand Haven reported to Little Point Sable to assist the ship. The ship rolled in the waves too far from shore to use the breaches buoy, too far to even read the name of the stricken vessel.

The ship being pounded by the waves was soon determined as a lifeboat, and a body of a sailor washed to the beach. The steamer in trouble was the *Novadoc*.

The *Novadoc*, a 235-foot freighter, had taken on a load of powdered coke and was steaming up the lake along the Michigan shore as the storm raged across Lake Michigan. The crew fought to keep the ship from being blown ashore. But despite all of their efforts, the winds gusting up to 80 miles-per-hour and waves as high as a house pushed the twelve year-old ship up on a reef off Jupiter Beach north of Pentwater. The ship grounded on the beach and being beaten by the waves, broke in two.

The Coastguardsmen made several attempts to rescue the nineteen crew of the *Novadoc* but each time the high wind and waves turned them back. The crew could only hold on while the waves smashed the grounded ship. Many of the crew were badly cut and bruised. Captain Steip received a laceration in his face and the first mate a deep gash in his cheek as part of the pilothouse was swept away.

In clear view of the hundreds of onlookers on shore, the crew of the *Novadoc* suffered their terrible ordeal for more than thirty six hours. They went without food, heat and in clothes soaked in cold Lake Michigan water. The crew shook uncontrollably from cold and feared that at any time a wave smashing down on the ship might be the one that would break it to pieces and send the men to an untimely death.

Two cooks who were trapped in the aft section of the ship were washed over and drowned, a fate the rest of the crew would soon share if not soon rescued. The ship was breaking under the strain of the continuous pounding of the waves.

Throughout the night, Clyde Cross of Pentwater, Michigan, paced the floors. He felt so helpless; there were sailors barely hanging onto their lives on the *Novadoc*, and he couldn't help. Clyde, one of nine brothers who made a living fishing out Pentwater, was respected for his skill and knowledge of the lake and by morning he had made up his mind to attempt a rescue of the seventeen men.

He asked Gustavo Fisher and Joe Fountain if they would volunteer to accompany him on his fishing tug to attempt the rescue. The two men were both men of the lake and knew when a life of a sailor was at risk other sailors did whatever they could to help and they quickly agreed.

The three boarded the small, old fishing tug *Three Brothers*, pulled on their rubber boots and rain slickers and set off from the safety of the harbor out into a raging lake. The wind was blowing a steady sixty miles-per-hour with gusts much higher, the snow blown before the wind so thick that visibility was greatly reduced and waves measured at thirty feet high crashed on the shore.

The scared and frozen men on the *Novadoc* saw through the blizzard the faint outline of a small boat heading in their direction. The boat rose on the crest of the waves and disappeared behind a wall of water only to reappear again drenched in a spray of cold Lake Michigan water.

Clyde Cross's boat, the *Three Brothers,* was considered by most as an old boat. The *Three Brothers* had long been a familiar vessel along the Lake Michigan coastline, involved in the thriving fishing industry. Many fishermen would not want to take the boat out on a calm lake much less Lake Michigan under a full blown gale, but Clyde Cross was determined to try to save the crew barely hanging on to life or die trying.

Slowly Captain Cross nudged his boat close to the *Novadoc* until he was close enough to throw them a line. The *Three Brothers* heaved on the waves and threatened to smash it against the large steel hulk of the steamer but Captain Cross skillfully maneuvered his boat close enough for the seventeen men of the *Novadoc* to jump from their precarious perch to the deck of the *Three Brothers*.

In minutes they were all aboard and the small fishing tug cut the line and backed away from the steamer. The rescue was not finished yet, they still had to navigate the treacherous waves and narrow entrance of the harbor. But in the experienced hands of Captain Cross the crew of the *Three Brothers* and *Novadoc* made it safely to shore.

During the Armistice Day storm of 1940 many individuals risked their property and their lives to help others in distress. From the courageous men of the Coast Guard stations of west Michigan to the heroic deed of Captain Clyde Cross, they did what they had to do for the benefit of others.

Results Of The Armistice Storm Of 1940

Lost

William B. Davock, 7,200 ton freighter, in Lake Michigan near Ludington, lost with the entire crew of 33.

Anna C. Minch, 4,200 ton grain carrier, in Lake Michigan near Ludington. Lost with the entire crew of 24.

The *Indian*, a fishing tug with five men from South Haven, Michigan.

The fishing tug *Richard H.*, with five men from South Haven, Michigan.

Grounded

The *Novadoc*, a 235-foot pulpwood carrier. Grounded and breaking apart on Jupiter Beach near Pentwater, Michigan.

City of Flint, Pere Marquette Railway car ferry (Number 18) with passengers and crew of 57 safe aboard.

Conneaut, a 415-foot freighter aground near Epoufette, Michigan off the Upper Peninsula. The crew is safe.

The gravel ship *Sinaloa* driven aground of Sac Bay. The crew of 42 were rescued by local fishermen and the Coast Guard.

The *Frank J. Peterson* was stranded on Hog Island. The crew were reported to not be in any danger.

"So Long Boys, And Good Luck."

The *Henry Steinbrenner* sailed into trouble in Lake Superior. Before the Coast Guard or other ships could arrive to render assistance, the crew had to fend for themselves. One crewman stepped forward and did what he had to do to save the lives of his fellow crewmen.

The 427-foot *Henry Steinbrenner*, owned by the Kingman Transit Company and operated by her namesake Henry Steinbrenner, was designed and built for carrying iron ore from the rich ore deposits in Minnesota and Michigan to the steel complexes in the lower lakes. The *Steinbrenner* had a long 52 year career on the lakes, but it was a career that was marred by a few "incidents."

The first of the problems for the *Steinbrenner* occurred in 1909 when she collided with the steamer *Harry A. Berwind*. The *Steinbrenner* was holed and sank to the bottom of the St. Marys River. Since winter was approaching the ship was left where she sank, on the bottom. She was later raised, repaired and went back to work on the lakes.

Twelve years later, in 1923 the *Steinbrenner* was again involved in a collision. This time it was in Lake Superior's Whitefish Bay with the vessel *John McCartney*. The *Steinbrenner* was damaged but made it safely to a shore facility. She was again taken out of service for repairs.

The next "incident" transpired in 1941. The *Steinbrenner* was again in the St. Marys River entering the Soo Locks when she rammed the lock wall and was so damaged that the ship had to be once again taken out of service.

The *Henry Steinbrenner* remained on the lakes until 1953 without any further significant "incidents."

On May 10, 1953 at 5:01 am the *Steinbrenner* departed Superior, Wisconsin with a cargo of iron ore. The iron ore was headed to feed the insatiable demands of the steel industry on Lake Erie.

When the ship departed, the weather was calm with no sea running. The forecast was for a south southwest wind to develop and reach a

The Steinbrenner *being raised after her 1909 sinking. From the Edwin Brown Collection of the State of Michigan Archives, Lansing, Michigan.*

velocity of 30-35 miles-per-hour. There was also a possibility for thunder squalls in the western portion of the lake. The weather was nothing that would hold the *Steinbrenner* in port.

The ship's cargo holds were equipped with hatch covers, each held down with twenty eight clamps. During heavy water, a canvas tarpaulin was fitted over the cargo opening before the hatch covers were clamped on. This step helped to keep the lake water from seeping in through the hatches.

The weather forecasts weren't calling for conditions so drastic that would require tarps, so they were left off.

About 3:00 pm, ten hours into the trip, the wind began to freshen and the waves increased. Captain Albert Stiglin, master of the *Steinbrenner*, radioed that the seas were building and they were taking green water over the bow.

The weather forecast was still predicting south southwest but the forecasters had increased the velocity slightly to 30-40 miles-per-hour, still nothing that would chase most Great Lakes freighters into shelter.

Crew was sent out to tighten the cargo hatch cover clamps. The order was also given to check and secure all deadlights and hawse pipe covers.

These orders were typical precautionary measures taken when the weather kicked up.

Around 8:00 on the evening of May 10 the storm worked one of the cargo hatch covers loose.

The third mate and three deckhands volunteered to go on deck and tighten the clamps. The ship was rolling in the seas and the waves breaking on the deck sent a spray up that could easily knock a man off his feet.

Since the conditions were so bad on deck the men wore a harness with a travel line connected to the ship's deck lifeline. The men braced themselves and walked in a wide stance trying to keep from being blown from the deck. Between the wind and the waves washing the deck, the task the men were to accomplish would be tedious at best.

At one point during the repair a particularly large wave broke over the ship and sent torrents of water rushing aft. The men were tightening the clamps when the wave struck. They could not help themselves from being washed away by the volume of water that poured aft along the deck. The men tumbled along the deck, thankful they were secured to the ship by the lifeline or they would have been washed over. If any of the

The Henry Steinbrenner. *From the Edwin Brown Collection of the State of Michigan Archives, Lansing, Michigan.*

men were washed off the deck into the sea, the conditions were so bad that the *Steinbrenner* could not have come about to rescue them.

As the men recovered from the wash that took them off their feet, they realized one of the men was missing. They worked their way back to the cargo hold and found the missing man, Tom Wells, hanging from his travel line in the cargo hold.

The horrible wind and huge waves tossing the ship as if it were a child's toy, the men struggled to pull on Tom's travel line, hauling him out of the hold and back on deck.

The men staggered on the rolling deck of the ship, aft to the galley/dining room. There they took time to regain their strength and were warmed by the cook's hot coffee. Once recuperated, the men went back out into the elements to fasten down the hatch cover clamps. All went back out except Tom Wells. He decided to remain in the galley, safe, warm and dry.

Conditions as bad as they were, they were about to get worse. Throughout the night the wind velocity increased, blowing as high as 80 miles-per-hour! The seas also continued to grow in height and intensity.

Early in the morning of May 11, the waves crashing down on the ship smashed in a door on the forecastle deck. Shortly thereafter the hatch cover secured by the crew had again worked loose.

With the winds blowing at up to 80 miles-per-hour and the waves running at 20- to 30-feet the conditions on deck had deteriorated to the extent that men could not be sent out to tighten the loose clamps.

The intense wind ripped at the hawse pipe covers and the clamps, loosening more hatch covers. Water poured in through all openings accumulating in the cargo hold faster than the pumps could remove it.

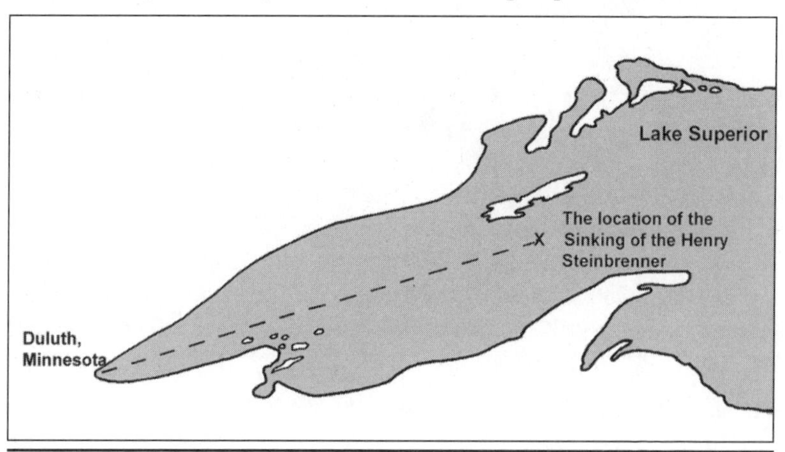

"So Long Boys, And Good Luck."

Captain Stiglin knew his ship was in trouble and decided to maneuver the *Steinbrenner* into a protected position between Isle Royal and Passage Island.

A little past 7:00 am on May 11, 1953 the captain gave orders to the crew to dress for the cold and put on their life jackets. He then put out a call to any vessels in the vicinity for assistance.

Just a half of an hour later the conditions aboard the *Henry Steinbrenner* worsened. The clamps on the three aft hatches loosened and the covers became dislodged. The mountainous waves broke on the bow of the ship and huge amounts of lake water ran the length of the ship and poured into the now open cargo holds at the aft of the ship.

"Stop the engines!" Captain Stiglin ordered. Then he sent out another message.

The message contained a request for any vessel in the area to come to their help and gave their location as 15 miles due south of the Isle Royal light.

The captain looked aft, and even through the wind blown spray and waves breaking on the ship, he could see his ship was going down by the stern.

The engines were stopped in preparation for the captain's next order.... Abandon ship!

The last sound a sailor wants to hear is the blaring of the abandon ship alarm.

As the harsh signal blared, the crew, without panic, prepared to abandon ship as they had so often practiced.

The ten crewmembers working at the forward end of the ship assembled on the forecastle deck at the life raft. At the stern the men gathered at the number one and number two lifeboats.

When the abandon ship alarm was sounded the captain also sent out a "Mayday" call to ships in the area and to the Coast Guard to come to their aid.

The ship was rapidly settling stern first as the waves broke over her and poured into the three open aft cargo holds.

The men at the stern readied the two twenty man lifeboats; lifeboat #1 was on the starboard, and on the port side lifeboat #2 hung from its davits.

The starboard lifeboat was swung out in a launch position away from the ship. But it prematurely lowered. The men held the painter, the bowline, but the waves carried the lifeboat away with only seven men aboard.

The D.G. Kerr. *From the H.C. Inches Collection of the Port Huron Museum.*

At the portside lifeboat, Third Assistant Engineer, Arthur Morse, helped other crewmen into the lifeboat. At launch Arthur refused to get into the boat, rather he elected to remain on the quickly sinking *Steinbrenner*. He knew the lifeboat would require help clearing the hull of the ship; he also knew it couldn't be done from the lifeboat.

As the lifeboat began to lower, a holding line that reduces the lifeboat swing and had not been disconnected in the rush to get off the sinking ship was discovered.

With the holding line attached the lifeboat could not be fully lowered and if the lifeboat remained attached to the ship, the *Steinbrenner* would pull the lifeboat and its passengers to the bottom with her. A quick thinking Art Morse pulled out his pocketknife and cut the line. Art looked at the men and said, "So long boys, and good luck."

That was the last anyone saw of Arthur Morse. He sacrificed his life for the life of his mates.

At the bow Captain Stiglin and nine men struggled to ready the six by twelve foot metal air tank and wood raft. The ship was pitching and rolling and waves broke over the deck hampering their efforts. A large wave crashed down at the bow of the *Steinbrenner*, the torrent of water lifted the raft and smashed it against the pilothouse wall, pinning Joe Radzewicz against the wall. Joe sustained a compound fracture to his left leg and broke his left arm.

Joe Radzewicz was no stranger to having a ship sink from beneath him. He had been aboard the *George M. Humphrey* in 1943 when, after a collision, the *Humphrey* sank in the Straights of Mackinaw.

As the ship sank lower, the ten men climbed into the raft and waited for it to float free of the ship. Just as the raft began to float a wave mounted the ship and capsized the raft; the crew were cast into the turbulent sea.

The men frantically swam towards the raft being tossed about on the waves. They struggled to climb back into the raft. The men were freezing and exhausted but they knew their only salvation was the raft. Swimming in cold Lake Superior water during a spring storm 15 miles from the closest land was a sure recipe for death.

Captain Stiglin and five of the men were able to climb back into the safety of the raft.

"...we lost four men for only six climbed back on the raft, including the injured Radzewicz." said Captain Stiglin.

The men could see the lifejackets of the four men but they were empty. Apparently the wicked sea stripped the men of their jackets.

The *Henry Steinbrenner* then slipped beneath the waves.

The men in the lifeboats and raft were off the sinking *Steinbrenner* but their safety was not guaranteed. They were drifting in a raging sea, with hurricane force winds, in sub freezing temperatures. To make matters worse, they didn't know if any ships had received their Mayday call or if they did, how close the ships were to their position or how long it would take for them to arrive on site. They wondered if the lake would take them before help arrived.

The Wilfred Sykes. *From the H.C. Inches Collection of the Port Huron Museum.*

The J.H. Thompson. *From the Hugh Clark Great Lakes Photographic Collection.*

Captain Stiglin's request for assistance and the Mayday were heard around the lake. Ships close to the coordinates the captain gave diverted their course and steamed to help. Lake men are quick to assist their brothers of the lakes.

The *D.M. Clemson*, *D. G. Kerr*, *Wilfred Sykes* and *Joseph H. Thompson* were all ships that responded to Captain Stiglin's call for help. The closest to the position of the sinking *Steinbrenner* was 13 miles.

The captain's frantic radio calls were also received by the Coast Guard and several vessels were sent following the *Steinbrenner's* Mayday.

Coast Guard rescue boats from Grand Marais, Two Harbors, Minnesota, Ashland, Wisconsin, Portage Canal, and Hancock, Michigan were dispatched. An air/and sea rescue airplane from Traverse City, Michigan, was also sent to aid in the search.

The men aboard the open boats, drenched by the 35 degree water, suffered terribly. The lifeboats and raft were tossed about by the 20-foot waves as if they were corks. The small boats rose on the crest of the waves but were then thrown into the trough and the men could see nothing but an angry Lake Superior in any direction. The wind, gusting up to 80 miles-per-hour, blew spray at such velocity it stung the faces of the terrified men.

Sitting low in the lifeboats and raft the men hoped they wouldn't be blown over. They clenched the thwarts and gunnels with ever numbing fingers, with thoughts of wives and children at home, and prayed.

About two hours into their ordeal the Coast Guard airplane was heard by the men. It raised their spirits but they knew the plane couldn't make a water landing in the tremendous seas. But they knew the plane would radio their position to other rescue vessels.

The men had floated helplessly on a wild Lake Superior for over four hours before Great Lakes freighters arrived on the scene.

The *Joseph Thompson*, the largest ship on the lakes in 1953, steamed into sight of the life raft. A rescue by such a large ship in such turbulent seas was going to tax the skills of the *Thompson's* captain and helmsman.

The huge ship made a head on approach towards the raft. A crewman aboard the ship threw a heaving line to the raft but the wind and seas caused it to miss its mark.

With great effort the big ship came about and made another approach. This time the heaving line was caught by deckhand James Lambaris aboard the raft.

"I held on for dear life." Lambaris said.

The bobbing raft was worked closer to the ship heaving in the heavy seas. The Jacobs ladder was lowered over the side of the ship to the raft, but the men were too numb and exhausted by exposure to the climactic conditions and shock to climb up to the deck of the ship.

The third mate of the *Thompson*, without thought of his own safety, descended the ladder swinging wildly and beating against the steel hull of the big ship. He tied lines around four of the survivors and they were

The United States Coast Guard Cutter Woodrush. *From the Richard Wicklund Collection.*

hoisted aboard the *Thompson*. Two other men on the raft needed to be raised to the *Thompson's* deck by use of a metal stretcher basket.

The Coast Guard boats and the other steamers patrolled the area searching for survivors. The steamer *Clemson* picked up the men, both alive and dead from one of the lifeboats. The ship took on seven survivors and four dead. The *Kerr* found and took on two bodies.

The *Sykes* came across the other lifeboat. The steamer maneuvered until it had the lifeboat along side. A line was attached from the lifeboat to the *Sykes*. A large wave caught the lifeboat and the line snapped. The steamer had to come about and work its way along the lifeboat again. The two survivors and one dead sailor were quickly removed from the lifeboat before the waves drove it away again.

The Coast Guard cutter *Woodrush* remained onsite overnight while the smaller Coast Guard boats went to shore. The steamers concluded their search and continued on to their destinations; the *Sykes* to Superior, Wisconsin, the *Thompson*, *Clemson* and *Kerr* to the locks at Sault Ste. Marie, Michigan.

The sinking of the *Henry Steinbrenner* in the early spring of 1953 brought the end to the 52 year old ship. Fourteen of her crew survived the trying ordeal, 10 bodies were recovered and seven men were reported missing and presumed dead.

On that day on the lake, there were many people who, at risk to life and limb, went out of their way for the lives of others. But no one gave more than Art Morse. The men of the *Steinbrenner* heaped mounds of praise on their shipmate Arthur Morse for his heroism and self sacrifice.

The Storm Of September 1930

The Great Lakes are notorious for storms which can produce hurricane force winds, blinding snow, and tremendous seas. One storm that assaulted the Great Lakes may not be as well known as the Great Storm of 1913 or the Armistice Day storm of 1940, but a storm that stands out in the annals of the lakes. It is the September 1930 storm that unleashed its fury on Lake Michigan.

The weather system approached from the southwest with terrible thunderstorms, severe lightning and 49 mile-per-hour winds. The storm raised havoc for nearly 24 hours before it passed over.

Most Severe Gale Recorded Locally Causes Much Lo[illegible]

Sweeping up from the southwest and accompanied by tor[illegible]rential rains and severe lightning, a 49-mile gale, the wo[illegible] ever recorded here, struck this district Friday mornin[illegible] causing untold loss on lake and land.

For nearly 24 hours the gale continued with hardly l[illegible]sened fury before it moved northeastward, toward Canad[illegible]

Strikes in All Quarters.

Orchards, farms, resort properties, city properties, li[illegible] of communication and shipping were heavily damaged.

The gale, whose velocity of 49 miles was equal to [illegible] under the former weather bureau system of measurem[illegible] was companioned by a 25-degree drop in temperature, c[illegible]ing householders to start grate and furnace fires.

HALF OF APPLE CROP IN COUNTY RUINED BY FURIOUS STORM

Appalling Damage Is Done to Fruit by Friday's Gale.

Is Worst Storm in History, as Far as It Affects Fruit, Says L. A. Hawley.

Total rainfall was [illegible] effectively ending the [illegible] drought and having one [illegible]ficial effect, that of [illegible] the forest fire menace.

Local shipping [illegible] home port Friday. [illegible] entered harbor shortly [illegible] noon Friday. Outbound [illegible] were held, three in L[illegible] one in Milwaukee and [illegible] Manitowoc, none leaving [illegible] 7 o'clock Saturday morning [illegible] regular schedule, Supt. [illegible] Mercereau stated yesterday [illegible] being resumed as rapidly as [illegible]sible.

Epworth Cottages Suffer

Epworth Heights resort [illegible] the full fury of the high [illegible] Portions of roofs of 38 c[illegible] were torn off and carpe[illegible] yesterday were busy repa[illegible] the damage. Trees suffered [illegible]

A headline from the Ludington Sunday Morning News.

Barns collapsed, roofs were ripped off, windows were blown in, trees uprooted and toppled, power poles snapped and farmers' crops were ruined.

On Lake Michigan the storm produced towering waves, reaching 30-feet high, or taller than a three-story building!

The fruit packet, *North Shore*, departed St. Joseph, Michigan with a cargo of 10,000 pounds of grapes bound for Milwaukee, Wisconsin.

The ship carried a crew of six including Captain E.J. Anderson, and his bride of two weeks.

TOLL OF FRIDAY'S STORM ON LAKE MICHIGAN NOW 11 DEATHS AND LOSS OF THREE VESSELS

Definitely Established Fruit Boat, North Shore, Went Down with Crew of Five Men and One Woman; Three Investigations in Salvor Wreck Indicated.

A headline from the Muskegon Chronicle.

The *North Shore* did not arrive in Milwaukee nor did she return to St. Joseph. It was hoped that the ship had made it to another port or had sought refuge in a deserted cove.

Hope for the *North Shore* was abandoned when large quantities of grape baskets washed up on shore.

The ship and the crew of the *North Shore* were never seen again.

Another victim of the September 1930 storm was the stone barge *Salvor*. The barge, equipped with a tall derrick shovel, drifted helplessly in the teeth of the storm after it broke loose from its tug north of Muskegon, Michigan.

As the storm shoved the drifting barge at will, seven of the crew took to the life raft. Several of them tied themselves in the raft to keep from being washed out by the waves.

Three men took their chances on the deck of the barge while two others of the crew climbed the derrick.

The men huddling on the deck were assaulted by the waves crashing over the barge and one by one were ripped from the *Salvor* and sent to a cold-water grave. The two men clinging to the girders of the derrick hung on as the storm raged, pushing the stone barge north until it grounded a distance from shore.

The Muskegon Coast Guard was notified that the *Salvor* had grounded and was being viciously assaulted by the storm whipped waves. The Guardsmen quickly pulled their shore apparatus to the beach near the grounded barge where they found wreckage littering the beach and bodies of several men washed up on shore.

The Harbor Beach Coast Guardsmen practice using the Breeches Buoy. From the Grice House Museum, Harbor Beach, Michigan.

The Coast Guardsmen quickly set up the Lyle gun and shot the line towards the disabled vessel. The purpose of the Lyle gun was to shoot a projectile attached to a light line to the vessel. The crew on the ship would then pull the light line, which was attached to a heavier line, to the ship. Once the heavy line was secured to the ship and to shore, a breeches buoy would be used to carry the crew to safety. The breeches buoy is a ring buoy hung horizontally from the heavy line. Attached to the ring buoy is what is best described as a rubber pair of pants. The men on the stricken ship would climb into the breeches buoy and the Coast Guard would pull them to along the heavy line to shore.

Unfortunately, all three attempts of the Coast Guard fell short of the stone barge *Salvor*. The ship was too far from shore to shoot a line to them with the wind blowing in from the southwest. Rescue of anyone left onboard would have to be made in a different manner.

Captain Preston and a crew from the Grand Haven Coast Guard Station set out into the storm at 4:00 am the following morning. The storm had not subsided and the Coast Guard crew in the open lifeboat met with difficulty in transit to the disabled stone barge. The winds had swung around to the west and the guardsmen took the full brunt of the storm on their beam for the entire two-hour trip to the *Salvor*.

When they arrived at the grounded barge, Captain Preston, at the tiller, skillfully navigated the lifeboat around the floating debris to get near the stricken craft. They looked for any survivors who may remain on the ship. Huge waves broke over the barge preventing the heroic men in the lifeboat from getting very close.

The Coast Guardsmen were about to give up hope when they saw movement from one of the men strapped in the derrick of the steam shovel. Knowing that there was at least one life to save, the Coast Guardsmen renewed their efforts. Unable to get close to the violently rocking barge Captain Preston yelled to the men in the derrick to jump into the water and they would pull them out.

The men, numbed by the cold, their 15-hour fight in the storm and sheer fright did as they were told and jumped. The water was churning all about them as they hit the water. They frantically swam towards the lifeboat and the heaving line the Coast Guard had thrown. The two men were rescued from an almost certain death.

Twenty miles off Ludington, Michigan the crew of another ship was fighting for their lives.

The ship was the 182-foot long, three-mast schooner, *Our Son*. Returning from Manitou Island with a load of pulpwood bound for the Central Paper Company of Muskegon, Michigan, the ship battled the huge waves. The schooner, not equipped with auxiliary power, was at the mercy of the winds in her sails.

The violent storm blew most of her sails to tattered shreds and the ship wallowed helplessly in the huge waves. Captain Fred Nelson, the 75-year-old master of the vessel, ordered the distress flag to be flown and hoped another ship would happen by and see their peril. The crew could do nothing but hope and pray.

Captain Nels Palmer of the Ludington, Michigan Coast Guard Station was notified by the Tenth District that they had received a wireless message from the steamer *William Nelson* that a schooner was in peril, forty miles southwest by west of Ludington. At 4:00 p.m. the captain and seven brave men from the Ludington Station set off in power surfboat to assist the schooner.

Hundreds of bystanders on the beach at Ludington looked on as the small boat powered out into the brunt of the storm. The surfboat was seen driving up on the waves then disappearing as it slid down into the trough.

Just in case of engine failure the masts had been erected on the surfboat, but the seven men onboard were not protected from the violent wind, crashing waves, and stinging cold spray. In the open water of Lake Michigan the Guardsmen could not set a course directly towards the coordinates the steamer had reported because the storm waves were

The old schooner Our Son. *From the Archives of the State of Michigan, Lansing, Michigan.*

coming from the south. Captain Palmer was lashed to the seat at the stern where he managed the tiller.

The surfboat was seaworthy and unsinkable but it offered the crew little protection from the wind and cold spray. There was no cabin and the canvas spray guard could not be used because of the high winds and the men suffered terribly from the cold and spray.

Seven hours into their voyage, at 11:00 p.m. the Guardsmen saw the lights of a steamer. Assuming it was the *William Nelson* standing by the disabled schooner, a course was set for it, but it was found to be another steamer plowing through the storm towards Chicago.

By taking a compass bearing on the reflection of the lights of Ludington and the reflection of Sheboygan, Wisconsin, the captain determined they were at the location where they should find the schooner. But it was not in sight.

They proceeded in a southwest direction for two more hours then to the northwest for three hours, but still there were ships to be found.

At 5:00 a.m., Captain Palmer decided to go to Sheboygan, Wisconsin, for gasoline, food and warm dry clothes. They arrived at the Coast Guard Station in Sheboygan fifteen hours after they had set out from Ludington!

The men had endured horrific conditions. They had not eaten, or slept during the ordeal and even though they wore boots and slickers the cold spray from the surfboat plowing into the huge waves soaked the crew to the skin.

Given dry clothes and food the heroic crew of the Ludington Coast Guard Station, shaking with cold as they sipped hot coffee, related the horrendous conditions they underwent the night before. They were preparing to head back out into the furry of the lake again when word came the crew of the schooner *Our Son* had been saved.

The *William Nelson* was heading to Indiana Harbor, Indiana, when in the distance she saw the schooner drifting in the storm, their American flag flying upside down as a distress signal.

Captain C. A. Mohr, master of the *Nelson*, at great risk to his own vessel, circled back to assist the schooner. The sea conditions were much too volatile for the steamer to attempt a rescue of the crew but sent a

The old schooner Our Son *after riding out the September 1930 storm. Note the American flag flying upside down as a signal of distress. The ship later sank to the bottom of Lake Michigan. From the State of Michigan Archives.*

CREW OF OUR SON TAKEN FROM HISTORIC SHIP IN THRILLING AND DARING SHOW OF SEAMANSHIP

Master of Nelson Makes Rescue at Risk of Own Boat as Waves Beat Craft

SCHOONER SINKS

Last of Sailing Vessels on Great Lakes Goes Down in Storm Still Fighting.

A headline from the Muskegon Chronicle.

wireless message to the Tenth District of the Coast Guard to report the position of the disabled vessel. Then again risking his own ship, Captain Mohr stood by the schooner in case the *Our Son* began to capsize and the crew had to abandon ship.

The storm was not abating and not sure when help would arrive for the crew of the old schooner, Captain Mohr decided to attempt a rescue of the sailors aboard the *Our Son*.

At great risk to his own ship and crew Captain Mohr maneuvered the *William Nelson* within hailing distance of the schooner. Sailors of the steamer yelled to the schooner to come into the wind and they would try to get close to the schooner to take them off.

Lifeboats were readied on the *Nelson* but launching a lifeboat in these conditions would be suicide. The only chance for the *Nelson* to rescue the crew of the *Our Son* was to nose up to her and have the crew jump aboard.

The waves crashed on the steamer, trying to push it off course as it inched nearer the drifting ship. A tremendous wave broke on the ship, ripping away one of the lifeboats and tossing it into the lake where it was smashed by the waves. At risk to his own ship, Captain Mohr slowly maneuvered the bow of the bounding steamer up to the schooner wallowing in the trough of the huge seas and with the help of the steamers crew the seven sailors of the schooner and her captain were able to jump to safety aboard the steamer. In keeping with the tradition of the Lakes, Captain Fred Nelson was the last to leave the ship.

As the *William Nelson* steamed away, the crew of the schooner watched their boat until it disappeared in the dark and stormy night.

The schooner *Our Son* went to the bottom of Lake Michigan on that 26th day of September in 1930. The *Our Son*, launched in 1875, was the last of the old-time sailing vessels. When she went down it ended the era of the sailing schooner carrying cargo on the Great Lakes.

But on Lake Michigan during one of the worst storms to assault the region, many brave men risked all for the lives of others.

THE SAND BEACH LIFE-SAVING CREW AND THE WRECK OF THE SCHOONER ST. CLAIR

On Lake Huron in the early days of October 1888 several people earned the status of "hero." Some others were given the title of "survivor" and unfortunately others will forever be known as "victims."

Weather on the Great Lakes in the fall of the year can be warm, sunny days and comfortable nights or days of black skies, violent winds, horrendous storms and giant waves.

Lake Huron's violent storms were the reason a protected harbor of refuge was built at Sand Beach, Michigan. Located 60 miles north of the mouth of the St. Clair River at Sarnia, Ontario, and Port Huron, Michigan, the harbor offered ships the opportunity to seek refuge from the lake before crossing the always-treacherous Saginaw Bay.

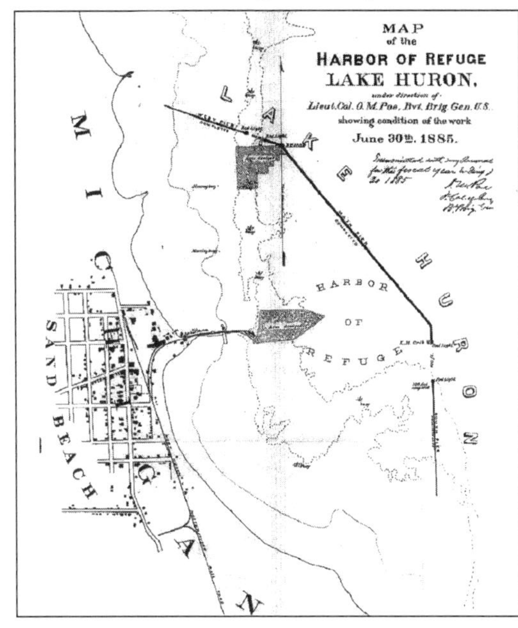

On October 1, 1888 the one-mile wide by three-mile long newly completed harbor of refuge was lined with steam and sailing vessels four and five deep.

A gale warning signal flag flew from the harbor master's lookout tower and the barometer was low, but some up bound ships left the safety of the harbor and headed out into the open waters of Lake Huron.

The 136-foot steamer *Lowell* with six barges in tow were amongst the ships which ventured out in the impending storm.

Many of the ships which left the harbor soon realized that conditions outside the safety of the harbor were worse than they had anticipated and came about and returned to the harbor.

The captain of the *Lowell* realized the northwest wind had reached gale force intensity and the waves were growing into small mountains. He decided that towing six barges through the storm was foolhardy and dangerous. The *Lowell* also came about to return to the protection the harbor at Sand Beach offered.

The *Lowell*, with black smoke bellowing from its stack. managed to get his string of barges heading south but the vessels were not safe yet. They still had to turn to starboard to enter the mouth of the harbor, a maneuver put the howling wind on their starboard quarter and a course which would put the *Lowell* and its barges in a precarious position while entering the narrow entrance of the harbor.

Sailors from ships already safely moored behind the breakwall watched as the *Lowell* steamed west towards the harbor entrance with the lighthouse flashing its inviting white and red signal.

As the *Lowell* neared the entrance, the towing hawser parted! The steamer was being forced precariously close to the breakwall. If the *Lowell* had continued on its course, with the gale blowing across its starboard bow and the six tows tugging at its stern, the steamer would

Ships sheltering in the Sand Beach Harbor of Refuge. From the David Busch Collection.

The brave men of the Life-Saving Service went out in any weather when the lives of Great Lakes sailors were in jeopardy. The photograph depicts a crew returning with the crew of a shipwreck. Author's collection.

have collided with the breakwall and the six barges would have smashed into the *Lowell*, the breakwall and grounded in the shallows and been destroyed by the waves.

The *Lowell* made it safely into the Government Harbor of Refuge but its six tow barges, old schooners and steamers were left in the open water to fend for themselves.

Crewmen aboard the barges quickly cut the towlines connecting them. If they remained connected to one another they all probably would have ended up breaking up on the rocky shore. Individually the ships had a chance at survival.

Each of the barges carried a captain and crew to assist in docking maneuvers, loading and unloading cargo and/or maintaining steerage while under tow. But, now the crews of the six barges needed to take command of their vessels which were drifting south on Lake Huron before gale force winds.

The crews of the barges *Seagull* and *Magnet* quickly raised sails and set a course down lake for the safety offered in the St. Clair River.

The other barges, the *Lilly May*, *St. Clair*, *Oliver Cromwell* and the *William Young*, one an old steamer which did not carry sails and the others too decrepit of condition to attempt to sail for the river, elected to lower their anchors and ride out the storm.

The captain of the *Lowell* enlisted the assistance of the tug *Champion* to go back out in the lake to try to round up his tows. The two steam vessels could not get close enough to the barges to render assistance due to the seas and shallow waters into which the tows had been pushed.

The Government Life-Saving crew at Sand Beach stood at the ready to render assistance to the ships laying at anchor. Once the *Lowell* and *Champion* returned to the harbor, the Captain George Plough, in command of the Sand Beach Station ordered his men to the 30-foot surf boat to go to the anchored boats.

The northwest gale blew the lake fury as the Life-Saving crew left the station and headed out into the lake. The eight men pulled at the 16-foot sweep oars as Captain Plough hugged the tiller. The waves broke over the boat and the wind drenched them with spray.

The crew battled the wind, waves and driving rain to the lee of the nearest barge, the *Lilly May*. The *Lilly May* was a sturdy vessel which seemed to be riding the storm well. Her anchor was holding and the captain declined the assistance of the Life-Saving crew.

The 138-foot *Oliver Cromwell* was built as the steamer *Dayton*, but after 35 years she was now just an un-powered barge. The anchor of the *Cromwell* was holding and the crew also elected to remain on board and ride out the storm.

Captain Plough directed the surfboat towards the schooner *St. Clair*. The *St. Clair* had been built in 1853 as a 153-foot long three mast schooner. The *St. Clair* was riding low in the water with a cargo of coal, her anchor chains taut with the strain.

Captain Jones and the crew of the *St. Clair* refused to abandon the ship.

Captain Plough pleaded with him to reconsider, but the crew were determined to ride out the storm on the *St. Clair*. Knowing the ship was of questionable seaworthiness, the Life-Saving Station surfboat remained on scene for about an hour but the rain and cold was freezing the crew and Captain Plough decided to return to the harbor. Before leaving Captain Plough again asked the crew to abandon the *St. Clair* but they again refused. Captain Plough told them to burn a signal torch if they changed their minds and wanted off the ship. The surfboat headed into the wind and pulled for the harbor arriving at the station at 9:30 PM.

The surfmen barely had time to change into dry clothes when at 10:00 PM a surfman on watch reported a torch burning on the *St. Clair*. The tired and weary surfmen again took to the surfboat to pull to the *St. Clair*.

Captain Plough later wrote in his report:

> "About 10:00 PM a torch was burned and we again started out. It was blowing harder and the sea was much

> heavier than upon our first trip. We passed the *Lilly May*. They again said they were alright. We then dropped down to the *St. Clair* and found all on board ready to leave her, and we got them into the lifeboat without any injury to themselves or damage to the boat."

The crews, consisting of Captain C. H. Jones, four sailors and the cook, a young woman, were off the *St. Clair* but they were not out of danger. Captain Plough knew they would need to pull into the gale to make it back into the harbor, a feat which would require near Herculean effort of his eight crewmen, so they made the decision to go with the wind and go down lake to Port Sanilac.

They rigged a small head sail and set off on what was going to be a wild 30 mile voyage. They sailed through the storm and driving rain with Captain Plough wrestling the tiller trying to maintain a course and keep the surfboat from broaching and rolling in the sea.

They were cold and tired but making good time until a wave broke over the stern and ripped the rudder from the boat. The crew, soaked to the skin, quickly used their oars for a rudder. Throughout the night they fought to keep the surfboat before the wind while waves broke over the boat.

The 17 year-old cook from the *St. Clair*, Julia Greaweath, sat in the bottom of the boat holding the lantern for Captain Plough to see the compass. The lantern was also a beacon for the people on shore following the progress of the surfboat.

While the surfboat fought seas through the night, the barge *St. Clair* dragged its anchor until it washed up on the rocky shore. The *St. Clair*, already in a decrepit condition was damaged beyond repair. When the anchor chain of the *Oliver Cromwell* broke, the captain steered the ship to shore and grounded her. The crew made it to shore in the ship's yawl.

Of the remaining two barges, the *Lilly May* rode through the storm and when the *William Young* was being blown towards shore they pulled their anchors and sails for the river.

After battling the seas all night, being drenched by cold waves and rain, the surfboat crew and their passengers were physically and mentally exhausted when the Port Sanilac lighthouse beacon was seen.

Spirits were high despite the conditions; the people on the surfboat, some shivering uncontrollably, knew their ordeal was soon to be over.

Crowds of onlookers braved the rain and cold to render what help they could. They cheered as the surfboat neared the Port Sanilac dock.

As the surfboat rounded the dock, the boat was broadside to the waves. The surfboat rolled violently in the trough until a larger wave flipped the boat and all of its passengers into the cold writhing lake.

The mate and a crewman from the *St. Clair* were able to cling to the overturned surfboat. Captain Jones, three sailors and the cook were too weak to fight any longer and were lost to the lake.

The nine member crew of the Sand Beach Life-Saving Station struggled to shore, some rescued by onlookers. The men were frozen and exhausted.

While the tragedy of the crew of the *St. Clair* was unfolding on the lake, the crew of another ship was in need of rescuing just a few miles south of the Sand Beach Harbor of Refuge.

When the *Mattawan*, a 143-foot steamer with the schooner *Gibraltar* in tow, left the St. Clair River the skies were sunny and the winds just a breeze, but the further north the ships traveled the skies turned dark and the wind blew into a gale.

As the conditions worsened the captain of the *Mattawan* decided to seek the shelter of the Harbor of Refuge. With the wind blowing a gale and the lake whipped into a fury, the captain knew entering the harbor while pulling a tow would be a challenge.

While off the harbor, the *Mattawan* suffered a mechanical problem.

Without propulsion, the steamer could not keep her bow into the wind and the waves shoved her into the trough of the waves. The ship wallowed in the trough of the waves, rolling violently with each pounding wave drifting down the lake. Not being able to control the steamer much less the schooner *Gibraltar*, the *Mattawan's* captain ordered a crewman to cut the towline and set the schooner loose. The *Gibraltar* had a better chance of survival on her own than tied to a steamer which was being pushed towards a rocky shore.

The *Gibraltar's* crew were unable to control the schooner and the vessel drifted ashore near White Rock, Michigan. The crew were able to board the ship's yawl and get safely to shore. The schooner was left to the destructive forces of the waves and was soon breaking apart.

The waves shoved the helpless *Mattawan* towards shore until she grounded on a reef. The steamer would soon be broken into pieces by the violent seas.

The reef was about three quarters of a mile off shore, but the waves continued to pound her at the stern, threatening to break the steamer to pieces and send the crew to an almost certain death.

They attempted to lower the ship's yawl but the waves crashing against the ship overturned the yawl before anyone could climb in it. A crewman donned two cork jackets and tried to right the yawl but he was washed away, fortunately he was found, near death, later several miles down the lake.

Two fishermen in the village of Forester, which is north of Port Huron, George and Harmon Allen, loaded an eighteen foot flat bottom wood skiff into a wagon to go to the aid of the steamer.

The Allen brothers were aware that the Sand Beach Life-Saving crew was busy with the crew of the *St. Clair*, so they knew the lives of the crew of the *Mattawan* were in their hands.

The Allen brothers were well known in the area for their sailing skills and fearlessness, but in the storm with winds howling greater than 50 miles-per-hour, in an open flat bottom skiff the brothers skills and determination would be challenged.

The horses pulled the wagon with the boat the six miles to the north near the stranded *Mattawan*. Following the wagon was a parade of people, some spectators, some eager to help the stranded crew of the *Mattawan*.

The group came to a stop along the shore almost abreast to the steamer. The ship, grounded on a reef almost a mile off shore, was flying a distress signal. It was being battered by the constant onslaught of the waves. The ship had already sustained damage and the waves threatened to rip her apart and send her crew into the angry seas. The large waves breaking on the rocky shore would almost certainly kill the sailors.

With the assistance of the crowd, the Allen brothers got the skiff out of the wagon and into the water. The crowd walked the boat and the two brothers out beyond the wave break and wished the Allens a safe return.

Harmon and George Allen had grown up on the shores of Lake Huron. Both sailed on the lakes becoming captains of several vessels before they returned to Forester, Michigan, to earn their living as fishermen.

Pulling at the oars, the brothers set a course for the rolling ship. The ship was laying with her bow to the southeast while the waves broke over her stern and washed along the deck. The torrents of water ran along the deck washing anything not tied or bolted down into the lake. The ship's spar had been broken and was laying across the deck with the wire stays and shrouds laying in a tangled heap.

They rowed to the lee of the *Mattawan's* bow where the ship shielded the skiff from the wind and waves. They weaved through the wire stays from the fallen mast and yelled to the crew on the ship.

Two crewmen met them near the bow but were reluctant to leave the *Mattawan* and entrust their lives to two men in nothing more than a skiff.

The brothers convinced them that they were out there to rescue them but if the crew did not leave immediately the Allens wouldn't be coming back. One man practically jumped into the small boat while the other slid down a stay into the skiff. The two brothers and two passengers were all the skiff could handle in the conditions. The Allens turned the small boat toward shore.

None of them were saved yet; they had to row the skiff with the waves mounting the stern, threatening to overturn the boat, then they had to maneuver the craft through the waves breaking on the rocks near shore.

Many of the gathered crowd waded out to meet the skiff. The two sailors jumped out and ran for shore but the Allen brothers asked the bystanders to turn them around and headed back out to the *Mattawan*.

George and Harmon Allen risked their lives that day to rescue the entire crew of the grounded steamer *Mattawan*. They made three dangerous trips out to the ship and returned with the frightened crew of the steamer.

On Lake Huron on that early day of October, 1888 several people earned the status of "hero." Some others were given the title of "survivor" and unfortunately others will forever be known as "victims."

Captain Mattison And D.L. Filer

The 159-foot *Tempest* steamed out of Buffalo Harbor on Lake Erie with two barges in tow. The barge *Interlaken* was destined for Toledo and *D.L. Filer* was filled with coal bound for Saugatuck, Michigan, on lower Lake Michigan, a journey which would cover a distance of over 800 miles.

The trip was nothing out of the ordinary for the old steamer, this 1916 season being her 44th year on the Great Lakes. She had seen a lot of miles, a lot of cargo and a lot of weather. The barges were the 23-year old *Interlaken* and the *D.L. Filer* was in her 45th season and had weathered her share on the lakes as well. The *Interlaken* and the *D.L. Filer* were built as schooners and sailed the lakes until steam vessels became more cost effective and rendered the old sailing craft of little use other than as a barge.

The *Interlaken* and *D.L. Filer* would spend the rest of their careers being loaded and off loaded and pulled from port to port. They would never again raise their sails, feel the acceleration as the sails filled, they would never again sail under their own power. Their era in history was passing.

In mid October, 1916 the ships sailed out into Lake Erie under clear skies, but there was a storm brewing. It was a storm which to this day is still called "Black Friday;" Friday October 20, 1916, when a severe weather system took aim at Lake Erie and unleashed its full fury on the ships and sailors.

Crossing Lake Erie, the *Tempest* towed the *D.L. Filer* towards Saugatuck. As the three vessels neared Toledo, Ohio, the *D.L. Filer* was anchored off Bar Point, near the entrance to the Detroit River, while the *Tempest* with the *Interlaken* in tow steamed into Toledo to discharge the barge and pick up another. The *Tempest* and her new barge would rendezvous with the *D.L. Filer* and both barges would be towed to Saugatuck.

An artist's concept of Captain Mattison and Oscar Johansson clinging to the rigging of the sunken D.L. Filer.

Unfortunately for Captain John Mattison, master of the *D.L. Filer* and its crew of six, the Black Friday storm was about to bore down on Lake Erie and the old schooner was in the way.

The schooner, anchored by the bow and stern, strained at its anchor chain as the winds increased and the waves began to grow. As the storm increased in intensity and the winds grew from a terrible 50 mile-per-hour wind to a hurricane strength 70 miles-per-hour, the *D.L. Filer's* anchor lost its bite and began to drag along the bottom.

About 8:00 pm, Captain Mattison called the crew together and yelled to them over the shrieking of the wind to be prepared for sudden action if it came necessary. Over the roar of the storm, the snapping of the forward anchor chain was heard.

The schooner was helpless in the storm. The *D.L. Filer* was blown off position before the wind. The ship dragged her stern anchor; the crew prayed it would grab.

The *D.L. Filer* was thrown about by the furious waves. With no engine or sails and no means of propulsion, the vessel was at the will of the angry lake. The ship was swept into the trough of the waves and rolled violently. Then an enormous wave pounded down on the ship, the force of the wave opened seams and smashed open hatches and companionways. Lake water poured in and the *D.L. Filer* began to settle to the bottom.

The conditions were so terrible the crew knew leaving the ship in the yawl meant certain death. One man ran to the foremast and started climbing the rope ladder. He knew Lake Erie was notoriously shallow and that if the *D.L. Filer* were to founder the upper masts might remain above the angry lake. The crew climbed as high as they could up the masts.

The six sailors climbed to the masthead as Captain Mattison remained on the deck. As the *D.L. Filer*, being beat by the seas and wind, slowly settled to the bottom, Captain Mattison took to the rope ladder.

Waist deep in the wind whipped lake, Captain Mattison struggled to climb higher, but the rope ladder came loose from its moorings from the weight of seven men clinging to it. The captain threw a leg around the mast and held fast to a rope loosely swinging near him. The rest of the crew were flung into the raging sea. They tried to swim back to the ship but all but one man were quickly blown out of sight.

Oscar Johansson, weighted down by his oilskins, with strength from some unknown source, struggled against the turmoil back to the mast. Now only Oscar and the captain remained, and they were not sure how long they could hang on.

The two men clung to the mast, the raging lake just a few feet below them, their weakening voices barely heard over the waves and storm. The men were freezing in the cold rain and winds; their fingers stiffening with cold. After three long torturous hours Oscar yelled to the captain, "Its all up with me, I can't hang on much longer!"

"Hang on Oscar!" the captain yelled. "This sea can't scare a Norseman!"

Oscar took a tighter grip and buried his face in his oilskins, trying to shield it from the biting wind blown spray.

Twice throughout the night Mr. Johansson began to slip but both times Captain Mattison was able to grab him and bring him back to the mast.

In the early hours of the morning the storm had subsided, the two men had been clinging to the foremast for almost nine hours. They were physically and mentally exhausted. The lights of a freighter were sighted, They yelled and waved but the ship was two miles away and the men weren't seen. Hope turned to despair.

Another freighter was seen but it too passed without seeing the two freezing men.

At a dock in Detroit, the D & C Navigation Company's liner, passenger ship *Western States*, was scheduled to depart the day the storm struck Lake Erie, but the company aware of conditions on the lake, ordered the ship held at Detroit. The following day the storm had subsided and the *Western States* was able to depart.

The *Western States* sailed south along the Detroit River nearing Lake Erie. Conditions were still rough but the worst of the Black Friday storm had passed. As the passenger ship rounded Bar Point, a lookout in the pilothouse spotted a strange sight. It was two masts of a sunken schooner. Captain Robinson was told of the sighting. He took up the binoculars and scanned the area. It looked like there were two men in the rigging. "Take it closer," Captain Robinson ordered the wheelsman.

Oscar Johansson and Captain Mattison, weakened from exhaustion and cold, were ready to give it up. The lake had all but defeated them. Both men were ready to cash it in, just slip into the lake and give up. Then the *Western States* came into view.

Captain Mattison wearily waved his cap to the ship and the *Western States* sounded its whistle, letting the men know they had been seen.

The big passenger ship slowly moved closer to the *D.L. Filer*, careful not to run aground on the Bar Point shoal. Stopping about 1,000-feet from the distressed men in the rigging, Captain Robinson ordered the ship anchored and a small boat lowered. The captain asked for volunteers to man the rescue boat. The weather was so bad and the seas so large that the crewmen would be risking their own lives in the attempted rescue.

Five crewmen quickly volunteered, they climbed aboard the yawl and were lowered down into the churning lake. The men pulled at the oars struggling to get to the men in the mast. Passengers on the ship watched the small boat and its five heroic occupants rise on the waves then disappear from view as the boat slipped into the trough of the wave.

Oscar Johansson, exhausted, and numbed from cold could no longer hold on and fell from the mast into the lake. Captain Mattison searched the lake for Oscar, but he never resurfaced.

As the yawl from the *Western States* pulled close to the sunken *D.L. Filer*, Captain Mattison jumped from the mast into the lake. The oilskins, or waterproof slicker he was wearing quickly filled and threatened to pull him down as he made a frantic attempt to swim towards the yawl.

The captain, half frozen and exhausted, was taken to the passenger ship. He was warmed by the stewards and taken to a stateroom, where he slept until the *Western States* arrived at Cleveland.

Lake Erie had destroyed the *D.L. Filer*, killed six of the crew, and left Captain Mattison to mourn for his men, but he had beaten the lake. He took all that the lake could throw at him and he survived.

Abigail Becker, The Heroine Of Long Point

Lake Erie, the fourth largest of the Great Lakes, is about 215 miles long and approximately 57 miles at it's widest. Major historic port cities grew along its shore such as Buffalo, New York, Erie, Pennsylvania, the Ohio cities of Ashtabula, Sandusky, Cleveland and Toledo and the Canadian cities of Port Colborne, Port Dover, Kingsville, and Port Stanley.

The lake played a very important role in the development of the Midwest of the North American continent. Immigrants by the thousands came across the Atlantic Ocean arriving at the docks in New York City. There they dispersed throughout the United States and Canada. Many took a train or boat via the Erie Canal to Buffalo and Port Colborne where they boarded ships to travel the length of Lake Erie to various ports on the lake. Many others sailed up through the Detroit River into the upper lakes to the ports cities Chicago, Green Bay, and Duluth and the many small villages and towns between.

Lake Erie, located geographically between Lake Ontario and the other of the Great Lakes, was the main thoroughfare for passengers but also it was responsible for the growth in the rich and vast farmland of the Midwest. The lake offered an inexpensive way to get the products from

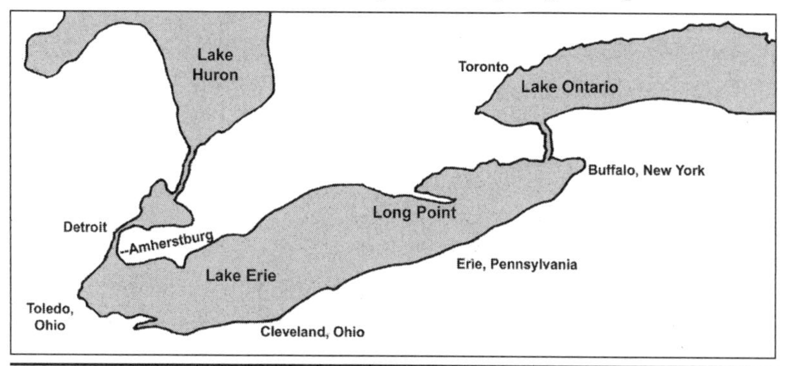

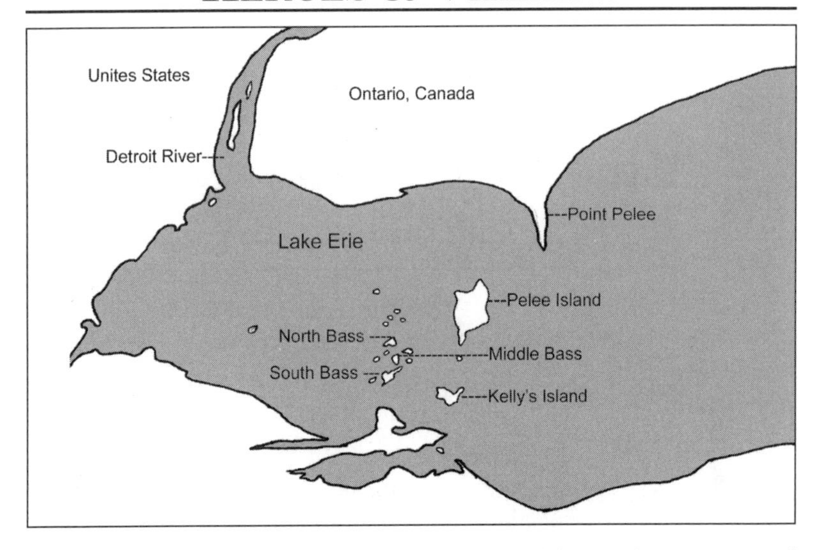

midwest farms to the thriving population of the cities on the east coast of the country.

At Lake Erie's western end, the Detroit River flows in. The lake waters then flow out of the eastern end at the Niagara escarpment where the lake water tumbles down a series of rapids and plummets down the Niagara Falls.

The most distinguishable feature of the western end of Lake Erie is the islands. Reaching up from the southern shore near Sandusky, Ohio, are Kelly's Island, the South, Middle and North Islands, the large Canadian Island of Pelee, and many other smaller islands. Point Pelee extends from the northern Canadian shore towards the islands creating a narrow deep water passage for the ships on the lake to navigate.

The most desirable feature of the middle to eastern portion of Lake Erie is Long Point. Long Point is a sandy peninsula, approximately 18 miles long, extending in a southeast direction out from the north shore of the lake.

Today Long Point is home to just a few hundred people but the population blossoms to over 5,000 when the summer residents move in. Summer cottages, golf courses, and campgrounds dot the peninsula, and a large tract of land is taken up by the Long Point National Wilderness Preserve, making Long Point one of southern Ontario's most popular summer destinations.

But it wasn't always a playground for urbanites. In the 1850s Long Point was an isolated, densely forested piece of land that was only inhabited by the most hearty hunters, fishermen and trappers. Jeremiah Becker was such a hearty soul who scraped out a living trapping and trading the pelts for supplies on which to live.

Jeremiah was a widower with a family of five boys and a girl. He met a seventeen year-old young woman, Abigail Jackson who loved the children and did not object to living in the wilderness of Long Point. They married and moved to the Point.

The couple and the six children lived in a small log cabin near the south shore of Long Point. Jeremiah would be gone for days on end checking his traps and taking his pelts across Inner Long Point Bay to Port Rowan. At Port Rowan, Jeremiah bartered his pelts for supplies the family needed to live throughout the winter.

On the morning of November 24, 1854, Jeremiah loaded his small open boat for the trip to Port Rowan. Abigail had an uneasy feeling about Jeremiah venturing out in the open bay. The weather was a bit breezy but otherwise a pleasant fall day. Just the same Abigail was not comfortable with her husband leaving. But in late November, you have to take the opportunity when it presents itself, for the fickle weather patterns of the Great Lakes change in a moment.

Abigail and the children watched Jeremiah set off on the lake until they could no longer follow his progress. They then busied themselves around the cabin. The southwest breeze and waves steadily increased throughout the day, the sky turned dark and foreboding and Abigail's thoughts were with Jeremiah.

That same morning at Amherstburg, Ontario, the schooner *Conductor* finished taking on a cargo of 10,000 bushels of wheat. The three-mast schooner was bound for Toronto, Ontario. The trip would take the ship through the Welland Canal from Lake Erie to Lake Ontario.

Captain Hackett decided to leave right after loading even though the sky to the west looked gray. He was afraid that since it was already so

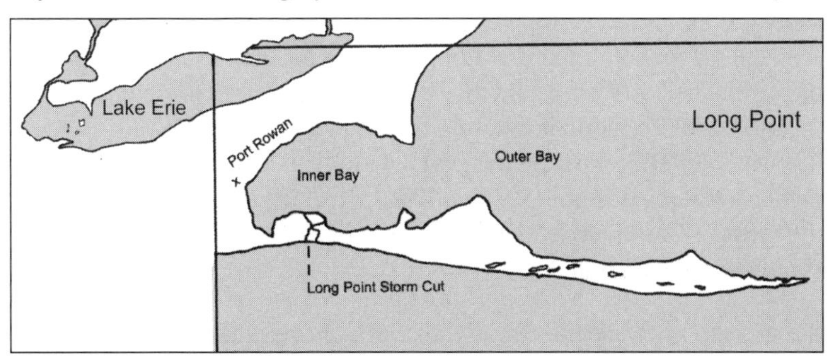

late in the season a winter storm might set in and freezing temperatures would close the locks along the canal. He wanted to get moving and get across Lake Erie, through the locks and up to Toronto.

The ship, with the captain and crew of seven, set off down the Detroit River into Lake Erie. Before long the storm from the southwest blew across the lake, winds grew to a high gale and built the waves to in excess of 12-feet.

Many years earlier a severe storm from the west blew the lake across the Long Point peninsula and created a storm cut, or natural canal leading from the west to the eastern side of Long Point. The navigable cut saved vessels several miles and many hours compared to sailing around Long Point.

Captain Hackett was sailing across northern Lake Erie, the wind and waves were tossing his ship at will and the blinding snow obscured visibility. The captain intended to sail for the Long Point Cut, pass through into Inner Long Point Bay and anchor in the lee of the peninsula, sheltered from the storm.

Without the ability to see any landmarks on shore due to the blizzard and the wind and waves tossing the ship about, the captain employed dead reckoning in hopes to make the cut.

Around midnight the topsail sheet was torn away and the sail quickly flogged itself to tatters, rendering the *Conductor* virtually uncontrollable. The ship was cast into the trough of the seas, rolling uncontrollably from side to side. The waves breaking down on her port side pounded on her deckhouse and eventually washed it and anything not secured overboard. The two yawls the ship carried were swinging in their davits, smashed against the ship and were reduced to kindling. The *Conductor* wallowed in the waves and the crew waited for the ship to capsize and throw them into the cold, boiling sea.

The crew endured the terror for over two hours when the ship suddenly ground to a stop. The ship had been pushed up on the outer sand bar off the south shore of Long Point. In the dark and blizzard snow, the sailors had no idea where they were. All they saw was darkness.

The waves breaking in the shallow water lifted the ship raising it off the sand bar. The ship with her port side to the weather was blown over on her starboard, the hull filled with water leaving the rigging just above the surface. The ship continued to be blown in her capsized state until the *Conductor* again grounded to a stop.

The captain and the crew scrambled into the rigging to save themselves. They lashed themselves to the rigging to keep from being blown into the tumultuous sea.

There was nothing the crew could do but hold on and pray they would make it through the night and be discovered at morning's light.

Throughout the night, Abigail Becker laid in bed listening to the howling wind and the roar of the waves breaking on the shore not far from her cabin. Abigail and her nine children, six from Jeremiah's first marriage and three they bore together, slept by the wood fire trying to keep warm. (Abigail Becker eventually raised seventeen children: six of Jeremiah's, nine of her own and two they adopted.)

In the morning Abigail stoked the fire and went to the lake for a pail of water to start the day. When she crested the dune she saw a ships yawl smashed on the shore. She saw no sailors and ventured down the beach to look for a ship in distress. After a short walk along the sandy beach she saw a schooner a mile or so from her position and several yards off shore.

She trudged along the shore towards the shipwreck, avoiding the waves breaking on shore and the snow blown into drifts. Once abreast the ship she could see the crew lashed in the rigging. Not knowing if they were still alive Abigail screamed, jumped and waved to the still figures in the rigging of the schooner.

There was no movement from the sailors. Then through the blowing snow she saw a slight movement. Was it a man weakly waving back to her or was it merely the wind causing the arm of a dead man to move? Then another man waved back to her. There were people alive out on the capsized schooner!

Abigail raced back to her cabin and gathered matches, tea and a teakettle. She awoke two of the older boys and had them follow her to the beach opposite the wreck.

The boys gathered driftwood on the beach and started a fire. The kettle was filled and placed in the fire. Meanwhile Abigail shouted to the men on the *Conductor* to jump into the water and swim for shore. They were weak and reluctant. Abigail walked out into the waves breaking at the beach. The waves pushed her back towards the shore and as a wave retreated from the beach the undertow threatened to suck her out into the lake.

She continued out until her sturdy frame, over six feet in height, was hip deep in the raging lake. She beckoned the crew to jump from the ship before the waves broke it up and sent them to a freezing death.

The fire blazing on shore drew the attention of the cold sailors on the ship and they yearned for the fire's warmth and the hot tea. Fearful of jumping in, the men feebly discussed their options. Remain onboard, pray the ship doesn't break up and wait until the storm subsides, or jump in and swim to the big woman and trust their survival to her.

Captain Hackett decided that if they remained much longer they would face certain death and told his men that he would put his life in the woman's hands and attempt to swim to her. If he succeeded then they should follow.

By now Abigail, who could not swim, had waded out until the wave surge was up to her chin. Captain Hackett unlashed himself from the rigging, removed his boots and slicker and jumped in. In his weakened state from being subjected to the wind, waves and snow throughout the night Captain Hackett frantically swam towards the woman. When almost within reach of Abigail the captain hardly able to move in the cold water and driving waves succumbed to exhaustion and the undertow began to suck him out to sea. Abigail reached out, grabbed the captain by his coat collar and pulled him free of Lake Erie's death grip to shore.

Captain Hackett was laid, completely spent, on the beach near the fire and given a cup of hot tea. Abigail turned back to the waves crashing down on the shore and waded back out for the next sailor.

Next to attempt to get to shore was the mate, John Jones. He shed himself of his rain gear and lowered himself down the halyard into the water. He made frantic swimming strokes until he was able to feel the sand of the bottom. He slowly walked along the bottom towards shore until the treacherous undertow swept him off his feet and began to drag him off shore.

Captain Hackett saw this and despite his exhausted state ran into the water to save his mate. As the captain neared John, the mate reached out and grabbed the captain in a death grip. Both men were now being washed out by the undertow.

Abigail ran as fast as the waist deep water would allow to the men. She reached out and mercifully grabbed an arm, then another. She turned towards shore dragging the two nearly drowned men.

Abigail, not questioning her fate, went into the raging water until all but the ship's cook remained lashed in the rigging of the *Conductor*.

The cook could not swim and despite Abigail's pleas for him to jump he would not. He feared dying by drowning more than he feared death by freezing. He wanted to stay and take his chances with the ship.

With the storm still raging without signs of abating, the weakened crew agreed there was nothing they could do. They left the cook lashed in the rigging of the foundered ship and slowly tramped to the Becker's cabin.

Throughout the night, the storm continued to assault the ship. The *Conductor* rolled violently with each wave that smashed down on the broken hulk while the cook slipped in and out of consciousness through the night.

At morning's light, the rescued crew members found the winds had abated some and built a makeshift raft out of logs and lumber washed up on shore by the storm. They paddled it out against the waves towards the remains of the schooner. They could see the lifeless form of the cook still in the rigging but they could not tell if he was alive or if the lake had claimed the poor soul of another sailor.

Upon reaching the ship, they yelled to him and saw a faint movement of his hand. More dead than alive, he had survived two nights tied to a grounded ship assaulted by everything Lake Erie could throw at it.

The cook was lowered to the raft and taken to the warmth of the Becker's cabin where he eventually regained his vigor.

In days, the storm had ceased and the crew, still suffering from exposure and frostbite, were sufficiently restored to make their way to the mainland. From there they were taken to their homes in Canada and Buffalo, New York.

The heroic deeds of Abigail Becker did not go un-noticed. The sailors she rescued told others of the "woman of Long Point" and how she rescued them from near death. Sailors took up a collection and presented a $500.00 purse to Mrs. Becker at a banquet given in her honor.

Since some of the crew of the *Conductor* were Americans from the Buffalo area, the Benevolent Life-Saving Association of New York rewarded her for her heroic deeds by awarding her a gold medal.

Mrs. Becker also was recognized in Canada by receiving a letter praising Abigail for her bravery from Queen Victoria. And later the Prince of Wales while touring Canada, personally visited Abigail and brought her gifts.

The story of how Abigail saved the crew of the *Conductor*, without any thought of her own safety, was talked about all over the Great Lakes. Sailors dubbed her the "Angel of Long Point."

The brave, heroic, "Angel of Long Point."

Right Time, Right Place To Be A Hero

Dr. Green, his wife and niece, Susan, sat having breakfast in the dining room of the passenger liner *Hamonic*. They had left Detroit just the day before, July 16, 1945 aboard the 349-foot ship with over 200 other passengers. Their journey would take them from Detroit up through Lake St. Clair and the St. Clair River, over the length of Lake Huron, then up the St. Marys River, through the famous Soo Locks into Lake Superior. Then they would return by the same course.

The *Hamonic*, now in her 36th season, was still a beautiful ship. In demand for a summer cruise, the Greens had to purchase the tickets more than four months earlier to reserve a cabin. The cabins were of a generous size. The dining room could rival any of the luxury ships sailing the Atlantic, and the ship offered many forms of shipboard entertainment.

The ship, having completed its first leg of the cruise was now tied up at the Point Edward wharf of the Canadian Steamship Lines just north of Sarnia, Ontario, across the St. Clair River from Port Huron, Michigan.

The Green family looked out the port side window at the United States side of the river and Port Huron. The river was busy with vessels of all sizes: kids in small home-built row boats, commercial fishing boats heading out into Lake Huron for their catch and the huge freighters passing with their loads of iron ore, wheat or coal. There were also the beautiful sailboats making practice runs in preparation for the Port Huron to Mackinaw race.

Susan, a typical 10-year-old, picked at her eggs and ham while watching the excitement on the river. Soon she went to the starboard windows to watch the stevedores loading supplies into the hold of the *Hamonic*.

"Look Uncle, there's some smoke coming from the dock. It looks like a fire." Looking out the window, they could see flames leaping from the top of one of the sheds on the dock.

The Hamonic *burns out of control in the St. Clair River. From the H.C. Inches Collection of the Port Huron Museum.*

The family went forward to the bow to better see the small fire, but within minutes, the fire jumped to other buildings and soon all of the sheds on the dock were ablaze.

The officers on the bridge watched as the blaze grew from the fire in the first shop. The First Officer signaled the engine room for power. He wanted to move the ship away from the wharf so fire and Coast Guard boats could get near to spray the flames, and to prevent an errant amber or flames blown by the wind from igniting the *Hamonic*.

The report came back that the boilers were down, and it would take some time to develop steam to move the ship. They would make haste in bringing up steam.

Susan and her aunt and uncle watched the building on the dock burn and collapse, thick black smoke rising above the dock, the slight wind blowing it towards the ship.

Rail cars, filled with supplies for the ship's week long cruise, were afire and several other boxcars nearby were in jeopardy of igniting.

Susan turned towards her uncle to ask a question, but her jaw dropped open as she stared over his shoulder in disbelief. "The boat is on fire!" she screamed.

Dr. Green turned to see flames leaping from the upper deck of the ship.

The burning Hamonic *loose in the river. From the H.C. Inches Collection of the Port Huron Museum.*

The ships of the day were constructed of steel hulls and main bulkheads, but most of the deckwork was made of wood. The *Hamonic's* wood deckwork, coated with highly flammable paint, ignited, supporting the flame and the fire quickly spread.

Passenger ships with long narrow hallways and many stairwells act as channels for the flames to spread. Within minutes the *Hamonic* was fully engulfed.

Captain Beaton raced for the bridge, pushing his way through groups of panicking passengers. The boilers had developed sufficient power to move the ship from the wharf, away from the fully engulfed dock. He yelled to deckhands to release the cables holding the ship to the wharf. Deckhands on the dock braved the heat of the fire as they pulled the heavy steel cables over the bollards and dropped them over the side. They ran for the bow, but the heat was too great for them to get near enough to release the ship.

The captain ordered the telegraph to reverse full, and hollered into the phone to the engine room to give it all they had. The huge propeller started to turn, churning the river into a frothing foam. The *Hamonic* backed off, ripping loose the mooring lines at the bow.

The crowds lining the Port Huron riverbank watched as the burning ship was backed into the middle of the St. Clair River.

"Full ahead!" Captain Beaton ordered the wheelsman. "But, captain?" the First Officer began in protest. Full ahead would run the ship into the river bank.

"The passengers have a better chance to survive if the ship is closer to shore," the captain responded, not waiting for the First Mate to finish his question.

The current of the St. Clair River is about 7 miles-per-hour. The captain knew many passengers jumping into the river would be swept downstream only to perish.

The telegraph clanged with the order, and the big ship slowed its backward movement, and began to move forward slowly, gradually gaining speed in the narrow river. With a jolt, the *Hamonic's* bow plowed almost 12-feet into the river's bank.

Doctor Green grabbed a life preserver and tied it around Susan. He looked for a way to escape. They ran to the nearest stairway leading below to the gangway but were met by a screaming group running up. "We can't get down! The smoke is too thick!" a passenger yelled to them.

They ran aft, but heat, smoke and fire blocked any chance for their escape. Dr. Green looked over the rail, thinking of jumping. He saw a small rowboat below. He lifted Susan over the rail and dropped her into the St. Clair River near a rowboat. He watched as Susan was pulled aboard the boat.

The Hamonic *was grounded into the river bank to save the passengers. The coal shovel can be seen off the port bow. From the H.C. Inches Collection of the Port Huron Museum.*

The Hamonic *burns while passengers watch from the safety of shore. The coal shovel which saved the lives of so many is seen removing passengers from the bow of the burning ship. From the archives of the* Sarnia Observer.

Dr. Green, content that his niece was safe, looked for an escape route for he and his wife, as did the other 200 passengers. Some were able to leave the vessel from the gangways before the fire grew too intense. Others elected to jump from the burning ship into the river, choosing the several story fall into the rapidly moving current of the river, where small craft rushing to their aid, picked them from the water.

Dr. Green and his wife saw people sliding down the heavy steel cables that once held the ship to the dock. It was their only chance for survival; they got in line to slide to safety.

The cables were hot from the intense heat of the fire, but burned hands were a small price to pay for your life. They slid down the cable until the pain was too great and let go, falling into the river near the riverbank where they were pulled to shore and taken to Sarnia General Hospital for care of their burns and smoke inhalation.

The ship lay almost perpendicular to the shore near the Century Coal Company docks. A quick thinking crane operator swung his bucket over to the bow of the ship. Passengers trapped on board ran for the bucket. Passengers climbed into the bucket, black with coal dust, for a ride to safety. Once his load was onshore, the crane operator raised the bucket up and quickly swung it over to the screaming people on the bow of the burning ship. The operator removed over one hundred frantic passengers from the ship 8 to 10 at a time in the coal bucket.

Vessels of all types worked the water around the burning *Hamonic*. Fishing boats, rowboats, race boats, anything that could be employed were used to pull the scared passengers from the river.

Harold Simpson, manager of the J. Wescott Marine Reporting Service, observed the ship catch fire from his office on the river. He and another man went to one of the company boats to help. While underway for the ship, they watched frightened passengers jump from the ship, a fall of 30 or more feet.

They came near a woman who held out her baby to them to be saved. Mr. Simpson found a towel to dry and bundle the baby, while the mother was lifted aboard. They continued picking passengers from the river. His boat full with survivors backed away to the screams of those still in the river. He begged them to move away from his boat so he could take his human cargo to shore. He promised to return.

Onlookers and other survivors of the ship ran to help unload Simpson's boat. "The river is alive with screaming people," Simpson said as they turned back for more.

A thick blanket of smoke hung over the surface of the river, adding to the panic of those floating in life preservers or hugging debris floating around the wreck. Screams of men, women and children haunted Simpson as the frantic passengers disappeared in the smoke. True to his word Simpson returned. In all he made four trips, taking more than 50 people to safety.

The ship's nurse, Dorothy Dure, was on the main deck helping frantic passengers over the rail to slide down one of the mooring cables. She spoke calmly to the frightened passengers, although, finding it sometimes necessary to raise her voice to keep the evacuation orderly. Standing not too far away she noticed a young boy looking all around and crying. He had become separated from his parents.

Dorothy went to him and reassured him that his parents were safe and that she would make sure he got to shore as well. She searched for a life preserver for him but all were taken. He stood by Dorothy until the last passenger had slid down the cable. Knowing the boy was not strong enough to hold onto the cable she had him climb onto her back for a piggy back ride. She stepped over the railing, tightly grasping the cable, and they began sliding down the cable to the water. The boy lost his grip on Dorothy and fell 15- to 20-feet to the river below. Dorothy splashed into the river and quickly swam to the surface to find the little boy. He was nowhere in sight. Dorothy enlisted the boats who went to her assistance to look for the boy. They looked but could not find him.

"Maybe he was picked up by another boat and is already on shore," one of the rescuers said. Dorothy, now frantic, searched the dock area but didn't find the boy. She reluctantly got in a car to be taken to the hospital. The boy was later found, wandering the wharf, crying, looking for his parents and the lady who helped him escape the fire.

In the following days, the extent of the damage became apparent. The fire was started by a faulty generator in the machine shop at the Canadian Steamship Lines' dock. All buildings at the wharf were destroyed as well as 30 railroad cars and their contents. The *Hamonic*, valued at $1,500,000, was beyond repair and salvage of the steel in her hull and machinery was the only recourse.

Through the bravery and heroics of many, the 220 passengers and 130 member crew were removed from the burning *Hamonic*, averting an almost certain catastrophe. One hundred and fifty people were taken to the hospital for burns, mostly to their hands and faces. Twenty-three were hospitalized, 13 in critical condition. Through the holocaust there was only one fatality, a deckhand who was killed as he fought to help others.

Keeper Captain Kiah And His Heroic Surfmen

The horrendous events which surround the heroic and tragic attempted rescue by the Port Aux Barques Life-Saving Station, have often been told. Yet it is an historical event that exemplifies the dedication and strength of the men of the Life-Saving Service.

April in Michigan's thumb area is beautiful. The cold of winter gives way to the sun and milder temperatures of spring. The trees bud and the fields start to turn green with growth. Ships can be seen on the lake transporting their first cargos of the new season. And the men of the United States Life-Saving Service begin practicing to be ready for any maritime disaster that might occur.

But Lake Huron does not give up winter's cold easily. The lake doesn't reach swimming temperature until late June, although some say it never does. Working on the lakes is wrought in inherent dangers. The spring, with its storms, cold winds, and near freezing water temperatures multiply the potential troubles. The worst of the dangers is the cold lake water. It will quickly rob a person of strength and reasoning. The life expectancy for a person in the icy water is short, very short.

On the cold spring morning of April 23, 1880 watchman James Nantau, of the Point Aux Barques Life-Saving Station, woke Captain Kiah just after daybreak to report he had sighted a ship in peril. The ship was about three miles distant and was flying its American flag upside down. A flag flown in such a manner is a universal a sign of distress.

The scow *J. H. Magruder*, loaded with 187,000-feet of lumber, was down-bound from Alcona for Detroit when a spring storm struck the lake. A spring storm on the Great Lakes usually means blinding snow, strong gusty winds and huge seas.

The ship was not far from the Point aux Barques lighthouse when the constant pounding of the waves parted seams of the ship's wooden hull. The *Magruder* began to leak and took on a bad list to port. Men were put to the pumps, but despite the constant pumping, 11-feet of water had filled her hull.

The ship, with a heavy cargo of lumber, was being blown towards the shallow. Captain Cronkey tried to control his ship but the weight of the cargo and of the water leaking into the hold rendered the *Magruder* sluggish and slow to respond to the wheel.

Captain Cronkey knew he could not sail to the Government Harbor of Refuge at Sand Beach with a heavy non-responsive ship, it would certainly roll in the seas. He chose to run for the shallows off Huron City.

The *Magruder* may have been in shallow water but not out of danger. She was laying bow to the east and taking the waves over the bow. Both port and starboard anchors were lowered in 14-feet of water, but with each surge of the waves the anchors dragged along the bottom. The ship, laying more than two miles from shore, was in peril. The crew needed to get off the ship before it was washed ashore and smashed on the rocks.

On board the schooner *Magruder* were the captain, four crewmen, the captain's wife and their two small children. They raised a red lantern in the rigging, and raised the American flag upside down at half-mast as signs of distress. Then they prayed someone would rescue them before the ship broke up on the reef and they were thrown into the cold waters of Lake Huron.

Captain Jerome Kiah, the keeper of the Life-Saving Station at Point aux Barques, saw the signal and ordered a surfman to light a Coston flare to tell the ship that their distress signal had been seen.

By 7:30 that morning the crew were prepared to go to the ship. Captain Kiah knew the waves breaking on the reef and rocks in the notoriously shallow area posed a challenge to the crew. He elected to use the station surfboat rather than the lifeboat for it was better in the breaking waves.

Onboard the boat was Captain Kiah, Robert Morrison of Caseville, William Sayres of Port Austin, James Pottinger from Huron City, Dennis Degan of Grindstone City and Walter Petherbridge and Jas Nantan, both from Walkerville.

The crew pushed the surfboat into the icy Lake Huron water and set out into the angry sea and howling east wind towards the *Magruder*, about three miles southeast of the station.

Captain Kiah at the stern with the steering oar and six surfmen manning the oars fought to keep the surfboat into the waves while maneuvering around rocks and avoiding the shallow reefs. After almost two hours they had plowed through the breakers on the reef into deeper water.

A painting by Robert McGreevy honoring the heroic attempt by Captain Kiah and the men of the Point Aux Barques Life-Saving Station to rescue the crew of the schooner Magruder. *Robert McGreevy, http://mcgreevy.com.*

Captain Cronkey master of the *Magruder*, watched the surfboat as it was struck by a tremendous sea and was tossed into the air, men and oars flying about. The surfmen were prepared for such an event, righting and bailing the surfboat was something they had relentlessly practiced. Captain Kiah was a task master when it came to drilling on the righting of the surfboat. It's a skill he told the men that someday might save your life.

The crew, floating in their cork jackets, quickly found their oars before they were washed away. As in the drill, they all gathered on the lee side of the upturned boat, hanging onto the boat's lifelines. They climbed on the gunnels, their weight forcing the boat to turn over. Once righted, the surfmen would crawl in carefully so as not to upset the boat and began to bail it out.

The men, soaking of cold April Lake Huron water, resumed their efforts towards the *Magruder*. They could see the ship's crew on the deck watching with anticipation of their rescue, the look of fear on their faces spurred on the Life-Saving crew. Captain Kiah was shouting words of encouragement to his men when another wave over took the surfboat and again it capsized.

The weary crew worked to right the boat, but this time they were exhausted. Every movement took effort the men no longer could muster.

Captain Kiah ordered the men to just hang onto the lifelines to regain strength before their next attempt to right the boat. Captain Cronkey, only a quarter mile away, could see the surfboat floating upside down, cresting high on the waves then dropping out of view in the trough. The exhausted crew hung onto the boat's lifelines.

Captain Kiah, realizing his crew was too exhausted to right the boat, told the men they would drift to the lee side of a point on the shore, then in calmer shallower water they could right, bail the boat and mount another attempt. Several times a wave broke over the boat scattering the men, but they were able to get back to the boat. Unfortunately, it took over an hour to reach the calmer water.

The captain and crew of the *Magruder*, fearful their vessel was going to soon break up from the constant onslaught from the waves, looked on in horror as their rescuers were tossed about by the huge seas.

Captain Kiah shouted over the roar of the waves and storm to encourage the men to hold out and reminded them of their wives and children depending on them.

Unfortunately, the cold was too much for them and the men one by one succumbed to hypothermia and slipped away into the icy cold water.

Men of the Point Aux Barques Life-Saving Station practicing near the Point Aux Barques lighthouse. From the author's collection.

The crew of the *Magruder* watched in horror but they were helpless to lend assistance.

A farmer, Samuel McFarland, alerted to the shore by his barking dogs, looked over the cliff and saw a small boat drifting towards shore. Unaware the Life-Saving boat was out he ran to the Life-Saving Service Station but found them gone. Mr. McFarland ran the distance to the Point aux Barques Lighthouse to notify Lightkeeper Shaw of the boat drifting towards shore.

The two men raced to the beach. The empty surfboat had grounded in the shallow water. They found Captain Kiah, more dead than not. He was standing, supported by a root from a fallen tree, incoherent, frozen, swollen, and blacken of face. He was making a walking motion yet his feet were not moving. He had spent over three and a half hours in cold, unforgiving Lake Huron.

The two men took the captain to the Life-Saving Station. They reported that he kept mumbling about his men and their bravery. The bodies of the surfmen drifted to shore and were found within a half mile from the station. Captain Kiah later credited the fact that his men were hot and exhausted from rowing and the cold water affected them more quickly than it did him.

Months later Captain Jerome G. Kiah and his crew were presented medals for their heroic attempts to save the lives of the crew of the *Magruder*.

Ironically, tragically and sadly, the *Magruder*, with its crew of men, women and children floated free of the reef with minimal damage. The six men froze to death and their captain was left physically and mentally depleted for nothing.

The Coast Guard Medal

On August 4, 1949, the United States Congress approved the establishment of the "Coast Guard Medal."

The medal is awarded to members of the United States Coast Guard who perform an act of voluntary heroism in the face of grave personal danger in such a manner that it stands out beyond normal expectations of their duties.

The gold medal is octagon in shape with the Coast Guard emblem in the center surrounded with a continuous rope.

The following are some heroes who have been awarded the distinguished Coast Guard Medal on the Great Lakes.

Rescue On The River

The Detroit River can be a fun waterway on a warm sunny day. Recreational boaters can take a short trip across the river to various Canadian riverfront restaurants, to the historic island of Belle Isle or head south on the river to many islands where on warm weekends hundreds of boaters party.

But the Detroit River can also be a dangerous waterway. The river is geographically located between Lake St. Clair and Lake Erie. Some of the dangers of the Detroit River are the current created by the volume of water flowing through the river and the large number of commercial ships on the river.

On May 21, of 1994 the weather was a windy cool 58 degrees, gray skies, with a chop on the river. At 2:00 pm a 21-foot recreational boat with seven passengers aboard capsized on the choppy river. Another boat nearby raced to the site of the accident and five people were taken aboard but two were missing; a 22 year-old woman and a 4 year-old boy.

An emergency radio call was made to the Cost Guard Station at Belle Isle. The crew responded in their rigid hull inflatable boat. When they arrived they found only the bow of the 21-foot boat floating above the water. The guardsmen could hear the pounding and screams of the victims trapped in the boat.

Chief Petty Officer Joe MacDonald, without thought of his own safety, stripped out of his clothes and leaped into the cold water in an attempt to rescue the pair trapped in the boat.

MacDonald, a Detroit police officer who served in the Coast Guard Reserves, reached into the boat for the victims to no avail. He then reached in with an oar but still the two inside did not or could not grab the oar to be pulled out. After 25 minutes in the cold spring water, which just a few weeks earlier was ice covered, Chief Petty Officer MacDonald climbed aboard the Harbor Master's boat that had just arrived with S.C.U.B.A. gear. A trained diver, Chief Petty Officer MacDonald did not take the time to put on a wet suit just the tanks, mask and fins. The lives of the two people trapped in the boat depended on him getting them out quickly.

He again entered the water, ignoring the cold and the fact that he was again exposing his bare body to the cold of the river. The visibility in the river was poor, slowing his progress and the boat's fishing lines and

6C THE DETROIT NEWS SUNDAY, MAY 22, 1994 ••

Woman drowns, five rescued when boat flips in Detroit River

other debris threatened to entrap him. The cold water was rapidly eating away at the chief's energy. He blindly groped in the forward cabin for the woman and boy. He grabbed the boy's leg and tugged with everything he had and pulled the boy to the surface. The four year-old was taken aboard a boat and rushed to Henry Ford Hospital.

Chief Petty Officer MacDonald again returned to the boat for the woman. The chief reached around the forward cabin hoping to touch the woman and pull her to safety, as he had been able to with the young boy. He stayed at the task until the cold water wouldn't allow his body to function any longer. He was pulled from the water after spending over 45 minutes in the cold water without a wetsuit or anything to protect him from the cold. He was rushed to the hospital suffering from hypothermia.

A member of the Wayne County Sheriff's dive team later recovered the lifeless body of the woman.

On April 21, 1995, Chief Petty Officer Joseph MacDonald was awarded the Coast Guard Medal. The Coast Guard wrote: "Chief Petty Officer Joe MacDonald demonstrated remarkable initiative, exceptional fortitude, and daring in spite of imminent personal danger in the rescue. His courage and devotion to duty are in keeping with the highest traditions of the United States Coast Guard."

EXTRAORDINARY HEROISM AT THE CLEVELAND HARBOR

A recreational sailboat on Lake Erie during a September storm can challenge a sailor's skills and endurance. On September 6, 1990, a small sailboat with six people on board was washed aground and broken up by 50 knot winds and huge waves near the breakwall at the entrance to Cleveland Harbor. All of the sailors were thrown into the wild lake, five were rescued but the sixth, a woman, was washed into the rocks of the breakwall.

A rescue boat from the Cleveland Coast Guard Station arrived on scene, joining an Ohio Department of Watercraft officer already there. With the fifty knot winds and the eight to ten foot waves, it was impossible to get a boat close enough to the breakwall to rescue the woman.

With thoughts of only the life of the woman trapped on the breakwall, Seaman Michael C. Pesce, volunteered to go to the breakwall to attempt a rescue.

The Coast Guard boat slowly nosed up to the breakwall and Seaman Pesce climbed out onto the rocks of the breakwall. Slowly he crept along the boulders illuminated only by his small flashlight and the flashing of lightening. His progress was hampered by the winds, driving rain and waves smashing and washing over the breakwall.

The waves crashed down on the seaman beating him into the rocks and twice washing him off the wall into the water of Lake Erie. He frantically swam back towards the breakwall. The beating he was taking and swimming for his life were taking a toll on his strength, but he kept going for the life of the woman trapped on the breakwall was in his hands.

When he reached her he discovered her legs were caught in a crevice in the rocks which held her underwater most of the time. It was just a matter of time and the woman would succumb to the relentless pounding of the surf.

Seaman Pesce repeatedly dove under the violent lake to free her legs as the mast of the sailboat and its rigging swung dangerously close to them with each crashing wave. Seaman Pesce's glasses were swept off his face and his helmet was washed away, but he continued to dive below the surface trying to save her.

Two other coast Guardsmen volunteered to go the breakwall and endured the horrible conditions to assist Seaman Pesce. Together the three men were able to pull the woman's legs free of the death grip the breakwall held on her. A rescue boat slowly eased up to the breakwall and the three Guardsmen lifted the woman aboard. There was not enough room on the boat for all four of them so Seaman Pesce remained clinging to a rock while the boat raced to shore with the woman in critical condition. Another rescue boat slowly edged to the breakwall and took an exhausted Seaman Michael Pesce off the rocks.

Unfortunately, the woman did not live through the frightful ordeal. For his bravery the United States Coast Guard awarded Seaman Pesce the Coast Guard Medal. The Coast Guard wrote; "His unselfish actions, courage, and unwavering devotion to duty reflect the highest credit upon himself and the United States Coast Guard."

SAVING A DROWNING MAN IN MANITOWOC

A Coast Guardsman isn't always on a mission to rescue victims at sea when they are called on to perform a heroic deed, sometimes they are called on when they are off duty. Such is the case of Boatswain's Mate First Class Eric Kyvik and Health Services Technician Third Class Grant Waldron.

The United States Coast Guard cutter *Mesquite* was berthed in Manitowoc, Wisconsin, and Boatswain's Mate Eric Kyvik was returning to the ship after leave. He noticed a commotion on the Eight Street Bridge. Several people and police were standing on the bridge looking over the railing towards the water.

Out of curiosity he went to the bridge to see what all of the commotion was about. He looked over the railing and saw twenty-five feet below a man floundering in the water. Without any thought to his own safety, only thinking of the man who needed help, Boatswain's Mate Kyvik, kicked off his boots, stripped off his coat and jumped over the railing into the Manitowoc River.

The U.S. Cutter Mesquite. *From the United States Coast Guard Historian's office.*

The cold water had taken its effect on the man and as he lost consciousness, Kyvik treaded water holding the man's head above the water to prevent him from drowning.

About that time Petty Officer Waldron happened by the activity on the bridge. Looking down he saw what was happening and quickly threw off his jacket and shoes and jumped over the railing to assist his shipmate. Landing one hundred yards from the two men in the water, Petty Officer Waldron quickly swam to them. He assisted in holding the man's head above the water while Mr. Kyvik swam pulling the men toward a boat coming to rescue them. The drowning man and the two Coast Guardsmen were pulled out of the cold river water and taken to a hospital where they were treated for hypothermia.

The two Coast Guardsmen, Boatswain's Mate First Class Eric Kyvik and Health Services Technician Third Class Grant Waldron, went to the aid of a person in trouble without a thought of their own safety. They jumped into the river to save the life of a man in dire straights. For their unselfish deed they both were awarded the Coast Guard Medal.

Introduction To The Villains Of The Great Lakes

Throughout history, many people have distinguished themselves for their heroic deeds. Men and women have disregarded the risks to their own safety to help a fellow man. But unfortunately, there have been just as many and probably more who have only risked their own lives for personal gain.

Criminals of the Great Lakes come in all shapes and sizes and their crimes range from those who during prohibition smuggled a bottle of booze across the border for their own personal use to those who could just as easily kill another than talk to him.

In between the extremes lie the many who have done their best to give the Great Lakes region a black eye. Unfortunately the information following covers only a few of the villains and criminal who have roamed the shores or sailed the Great Lakes. The Great Lakes have endured many villains; too many for one volume to categorize.

The King And The Great Lakes

In 1844, James Jesse Strang, a man of a variety of occupations; a postmaster, a lawyer and a Baptist Minister, moved with his family to Nauvoo, Illinois.

In Illinois he became involved with a religious organization which was in its infancy, the Church of the Later Day Saints, the Mormons.

James Strang met the church's charismatic leader, Joseph Smith, and the two became friends. In their conversations, Smith told Strang about the ideology of the church and within a year Strang had joined the church and was baptized. In just a few months James Strang was named an elder of the church.

While in jail for vandalizing a newspaper office that had written critical articles about the church, Joseph Smith was killed. The church was thrown into turmoil. Two elders of the church, Brigham Young and James Strang, claimed to have been Joseph Smith's chosen one to lead the church.

Members of the church chose sides with the majority siding with Brigham Young. Strang

Joseph Smith the charismatic founder of the Mormon Church. From the Church of Jesus Christ of the Latter Day Saints.

Brigham Young opposed James Strang in assuming the leadership of the Mormon Church after the death of the church founder Joseph Smith. From the Church of Jesus Christ of the Latter Day Saints.

and his followers, called Strangites, were excommunicated from the church and Brigham Young led his followers west to Utah.

Strang and his followers moved to Voree, Wisconsin (later re-named Spring Prairie). In Voree, the Strangites were harassed for their beliefs which differed widely from that of the predominately Lutheran population.

Strang told his followers that he had received divine revelations from God and that he was indeed a true "Prophet and Seer of God." He used his new Prophet status to rule his followers.

Strang issued new church edicts claiming that all material possessions of the Strangites belonged to the church, he would not permit the eating of red meat, and he demanded his followers to maintain strict sexual standards.

Jesse Strang preached that polygamy, or multiple wives was a sin and was not to be practiced in his church.

As the non-Mormon citizens of Voree began to question the religious beliefs and the strangle hold its leader held over his followers, Strang decided it was time to move to a location where they could practice their religion without persecution.

The leader and his flock first looked at Charlevoix in northern Michigan. The area was on Lake Michigan, there were abundant woods and its rural setting would provide the anonymity he sought. The area seemed almost ideal until James Strang discovered Beaver Island.

The island located in northern Lake Michigan offered a natural protected harbor, thick forests and abundant lake fish. The island was sparsely populated by just a few families of fishermen. With less people to interfere with James Strang and his followers, Strang selected the island as the home for his church.

A photograph of "King" James Strang. From the Church of Jesus Christ of the Latter Day Saints.

By the close of shipping during the winter of 1848, over 125 members had moved to the island. Within a few years the number had more than doubled.

The followers of Strang blindly obeyed him doing whatever he asked of them and they believed whatever he said. On July 8, 1850 James Strang told his followers that God had spoken to him and instructed him to declare himself to be the "King of the Kingdom of God on Earth".

"King Strang," with his new authority, began to make changes on the island which was now referred to as his kingdom. He established a tax levied on all non-church members who lived on the island. Many islanders were outraged and refused to pay the tax but relented after several men were beaten for not following the "King's" demands. The "King" also was accused of taking non-church members property for the church's use. His attitude was that if the Kingdom of God needed something, he could take it.

Strang had his members run for political office, allowing the "King" to extend his authority throughout the county. With his members in political positions of authority supporting him, Strang ordered the county treasurer to pay him a percentage of the county's tax assessment.

"King Strang" did improve the island and the lives of the Strangites by building a saw mill to provide the necessary materials to build a large church, homes and other structures for his growing membership. He established a newspaper, The Northern Islander, to promote his religious beliefs. He also printed books and pamphlets expounding his church's ideology and other propaganda.

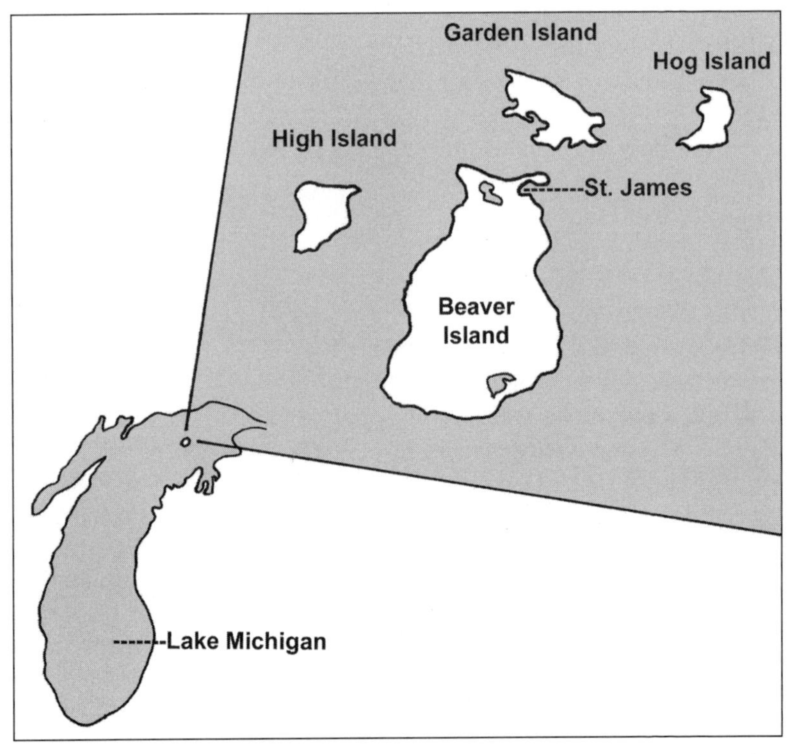

As the tyrannical "King" tightened his grip on his followers, he also tried to force his thoughts and beliefs on the non-church citizens of the island. Despite his earlier opposition to polygamy the "King" now embraced it and took a second wife. But, it was the King's decision to ban whiskey on the island that caused the most friction between the church and the non- church islanders.

The whiskey ban resulted in several altercations between the two groups. The confrontation came to a head when several drunken fishermen upset with the whiskey ban marched on the church threatening to throw the Strangites off the island.

As the angry mob approached the church, the Strangites fired a cannon at the gang of unruly islanders. The islanders dispersed but they began to leave their island homes for safer residences on the mainland of Michigan.

This was the beginning of the end for "King Strang." His confrontations with non-church islanders and his increasingly tyrannical behavior caused unrest among some of the Strangites.

The activities of "King Strang" did not go unnoticed by officials of the State of Michigan. They requested that President Millard Fillmore do something about this man passing himself off as a "king" on American soil.

President Fillmore was furious with the renegade religious leader and ordered the United States Attorney General to investigate "King Strang."

The President soon ordered the United States Navy Gunship *U.S.S. Michigan* to Beaver Island to restore peace and order.

The Navy Gunship *U.S.S. Michigan* was a historical ship. The ship, launched in 1843, was the first iron-hulled warship in the United States Navy. The 163-foot ship was originally designed as a three mast topsail schooner but was changed to a paddle sloop, a sailing sloop with a steam engine which powered a paddlewheel.

The *U.S.S. Michigan* was launched at Erie, Pennsylvania for use on the Great Lakes. The ship was involved in putting down civil unrest in port towns, battling lumber pirates, acting as a rescue vessel for ships in distress, and intervening in violations of federal laws.

In the spring of 1851, the *U.S.S. Michigan* entered James Harbor on Beaver Island with cannons loaded, sailors armed and ready for battle.

The U.S.S. Michigan *was sent to Beaver Island to quell the disturbance and to bring "King" Strang to justice. From the United States Coast Guard Historical office.*

DAILY FREE PRESS

S. M. JOHNSON AND T. F. BRODHEAD, EDITORS.

THURSDAY MORNING JULY 10.

The Mormons Acquitted

The trial of James J. Strang, and other of the Beaver Island Mormons, on an indictment for obstructing the U. S. Mail, which has been for some ten days past pending in the U. S. District Court, was yesterday concluded, by the acquittal of every defendant. The cause was submitted to the Jury on Tuesday evening, with instructions to render a sealed verdict. The verdict was agreed upon without hesitation by the Jury, who were but a few moments in consultation.

Also on board were U.S. Marshals, Sheriff's Deputies and a United States District Attorney. "King Strang" and about one hundred of his followers were taken in custody aboard the ship and transported to Detroit to stand charges of treason, counterfeiting and trespassing on federal land.

James Strang being a lawyer and a man of enormous ego, elected to defend himself of the accusations in federal court and won the case. He was found innocent of all charges.

James Strang returned to Beaver Island with his ego larger than ever and his authority even more strong. Based on the publicity of his trial, James Strang ran for and was elected to the Michigan State Legislature in 1853.

"King Strang's" reign began to falter when he accused a Strangite, Thomas Bedford, of adultery. "King Strang" ordered that he be flogged. Strang then excommunicated another Strangite, Doctor McCullough, for drunkenness. The two men held bitter contempt for Strang and conspired to kill the "King."

On June 16, 1856 Bedford shot James Strang three times. One bullet grazed his head, the second struck him in the cheek and the third lodged in his spine. The "King" was transported back to Voree, Wisconsin, where three weeks later he died from his injuries at age 43.

A group of non-church members from Mackinaw and other locations around Michigan and Wisconsin went to Beaver Island and forcefully evicted the approximately 2,600 Strangites still on the island. They were put on steamers, schooners and any other vessel available and deposited to various Lake Michigan ports.

Many of the fishermen who were forced to leave Beaver Island when "King Strang" took control returned to the island to find it vastly improved. What was once a small, very rustic fishing community now had paved roads, many new buildings, new homes, and improved docks.

The end of "King Strang" ended the loosely connected division of the Mormon Church. The egotistical leader, James Strang, had lead his followers from Illinois to Wisconsin to a small island in northern Lake Michigan where he created his own Nirvana. But, the "Kings" ego and huge appetite for power and control was his downfall.

Today Beaver Island is a peaceful island with several year-round residents and many summer cottage folks. A trip to the island is strongly recommended to enjoy Beaver Island and its history.

Pirates On The Great Lakes

The word pirates brings to mind a swash buckling, eye patched, peg legged, hook hand individual who freely roams the sea and takes whatever he wants. While this description was popularized in Robert Louis Stevenson's *Treasure Island* and perpetuated in Hollywood, it is far from the reality of a Great Lakes pirate.

Pirating on the Great Lakes was not quite as glamorous as the image the Hollywood screenwriters have created. Both the ocean going and the Great Lakes version were usually low life criminals who practiced their illegal deeds on the water, in addition to their shore side criminal acts.

The Great Lake criminals committed many forms of piracy. Cargo was stolen, ships were seized, and violent acts were committed against ships and crews.

When the early explorers first ventured into the Great Lakes region, they were impressed with the abundance and variety of animals. They knew the animals could be trapped for their pelts and sold in Europe for a handsome profit.

From 1600 until 1760, the French held exclusive rights over the region and the animal pelts therein. They traded blankets, iron tools and cotton cloth to the Native Americans for furs they had trapped. The French established trading posts on Lake Superior where beaver and other quality pelts trapped by Natives were exchanged for the manufactured goods shipped from Europe.

From 1754 until 1763, the English and French were at war in Europe. In America and on other continents, they both claimed territories. The war in the America's is called the French/Indian War because the English were actually at war with France and its Native American allies with whom they had established a relationship through trade.

To help defeat the French, the English employed a tactic often used in previous wars; they hired privateers.

A privateer is a pirate authorized by the government of a country to attack another country's ships. Not quite a pirate who attacks and seizes ships randomly, but a person who is under contract by a country to perform acts of piracy on the ships of its enemy.

A pirate kept the ships and treasure they obtained for themselves and a privateer turned over the ships and treasure to the contracting country and received a commission.

The English government commissioned George Colby to be a privateer. Although Colby and his men did not attack French ships from large schooners or brigantines as Jack Sparrow did in the *Pirates of the Caribbean* movies, they instead harassed the French ships from small boats and from shore.

One of their tactics was to build fires at night on the shore of Lake Erie to confuse the French into thinking the fire was a signal fire indicating they were near a port. The ship would change course to enter the port but rather than finding the city, they would end up on a shallow

The logs floated down stream from the forests to the mill and were often pirated by unscrupulous loggers. They guided them into side streams, sawed off the company's mark and were restamped with a new company mark. From the Bayliss Public Library, Sault Ste. Marie, Michigan.

A log end showing the lumber company's mark indented into the wood. From the author's collection.

rocky reef. The ship and its cargo was destroyed and Colby and his men would attack the French ship. George Colby's act of piracy on the Great Lakes helped the English defeat the French.

In 1760 the English had conquered the French and declared all of the French territory as belonging to them. All pelts were now sent to London instead of Paris and the goods traded with the Native Americans now originated in England.

The English set up trading posts for the Native Americans to deliver their pelts and trade of goods.

At the English trading post on Drummond Island, an unscrupulous post commander made it a policy that all of the Native Americans, men, women and children must have a drink of rum, or several, as a display of friendship. The drinks were consumed before negotiating what would be received for pelts. The Native Americans would leave the trading post with little or nothing to show for trading their pelts. They relied on trading the pelts for food for the winter but all they had were small amounts of food and worthless trinkets. While this was not an act of piracy it was a villainous act!

An axe with a lumber company's identifying brand on the flat side. The axe would be swung at the end of the log, stamping the mark into the log end. From the author's collection.

After the fur trade, the next exploitation of the region was for its huge lumber resources. Most European countries had been deforested by the demand for lumber but in the Great Lakes they found huge trees growing to the sky and forests so thick one couldn't walk through them. The majority of the people populating the United States in the 1800s were settled along the eastern coast. The lumber they needed to build their homes and other buildings came from the forests close to the cities but they were being logged out.

By the 1860s, the eastern forests were depleted and the Lumber Barons looked to the virgin growth forests of the Great Lakes area. Michigan was ideal for their needs as it had vast forests of virgin hardwoods and white pine, most of the state's terrain was gentle, there were plenty of rivers to transport the logs to saw mills and the Great Lakes offered an inexpensive way to get the lumber to the cities.

The method of logging in those days was for the Shanty Boys, the men who cut down the trees, to use axes or hand saws to down the trees during the winter. The logs were pulled across the frozen ground to the river's edge by mules and oxen by Teamsters. When the ice and snow melted in the spring, the rivers and streams swelled to several times their normal size, men called River Hogs pushed the logs into the rivers to be floated down stream to the saw mills.

This procedure of floating the logs to the sawmills created an opportunity for some nefarious scoundrels to acts of piracy.

To prove ownership of a log, and to direct it to the correct mill, or credit it to the correct company, the Shanty Boys stamped an identifying mark, a company mark, on the end of the log. They used an axe with the brand of the lumber company on the flat end opposite the blade. With a mighty swing the axe striking the log end indented the company's mark in the wood.

Some unscrupulous loggers pirated or redirected the best logs floating downstream into side streams or hidden ponds. There they cut off the branded end of the logs and re-stamped the log with their own identifying brand. The logs were then put back into the river and floated to the mill where the thief was given credit for the log.

In the 1800s travel around the Great Lakes by ship was the main form of transportation. There were few roads, and those which did exist, were often impassable. The railroads only connected major cities, and air travel was only a dream in some "foolhardy" individuals. If passengers, cargo or livestock needed to be transported, ships were often the only option.

At times, a ship in upper Michigan or Wisconsin was loaded with lumber in preparation for transportation to Chicago, Detroit, Cleveland or another populated area. But, the next morning when the crew boarded to depart they found that much of the cargo had been stolen during the night. Thieves re-loaded the cargo on their own boats and sailed off to sell it in the major markets as their own.

Some criminal lumberjacks would cut huge tracks of forest from Great Lakes islands, or from desolate locations from land that belonged to the federal government. They would then load their ships with the logs or lumber and sell it in the larger cities. This practice became so prevalent in the 1850s that the United States government passed a law protecting the tracks of timber held by the government for the Navy's use. The trees were needed for building ships for the Navy. The tall straight trees were perfect for the keel, masts and planking of wood ships.

To combat the timber pirates, the Navy's first iron-hulled ship, the *U.S.S. Michigan*, was assigned to put an end to the piracy.

The ships used to transport the illegally gotten lumber were no match for the *U.S.S. Michigan* and the timber pirates were soon put out of business.

Pirates And Lake Erie's Johnson's Island

Today the American Civil War is considered a war fought in the southern regions of the United States. Yet in 1864, the war came to the Great Lakes.

In early battles, the Union states succeeded in capturing many Confederate prisoners of war, so many in fact that housing them had become a problem. To deal with the problem, the Lieutenant William Hoffman, Commissioner General of Prisoners, received authorization to explore northern locations to serve as a prison for Confederate soldiers.

He determined an island in the Great Lakes would offer the most secure setting. An island would prevent prisoners from escaping and an island would offer protection from the enemy trying to invade and free the prisoners.

Lieutenant Hoffman looked to the islands along the southern shore of Lake Erie as possible locations for a incarceration camp. The area offered several possibilities, North, Middle and South Bass islands, Pelee Island and Kelly's Island were considered.

Pelee Island, the largest of the islands, was eliminated because it is a holding of Canada and the Canadians were sympathetic to the Confederate states and would never agree to a Union prison being built there.

Middle and South Bass Islands met the requirements Lieutenant Hoffman had established but they were eliminated due to their proximity to Canada and the Union's inability to adequately protect the outpost.

Kelly's Island was considered but it was eliminated for many of the same reasons as the Bass Islands. Lieutenant Hoffman decided to build the prison on Johnson's Island.

The 300 acre Johnson's Island is located a short distance from the Marblehead Peninsula in Sandusky Harbor. It was close enough to easily protect from Confederate attacks and supplies could be transported to the island around the year.

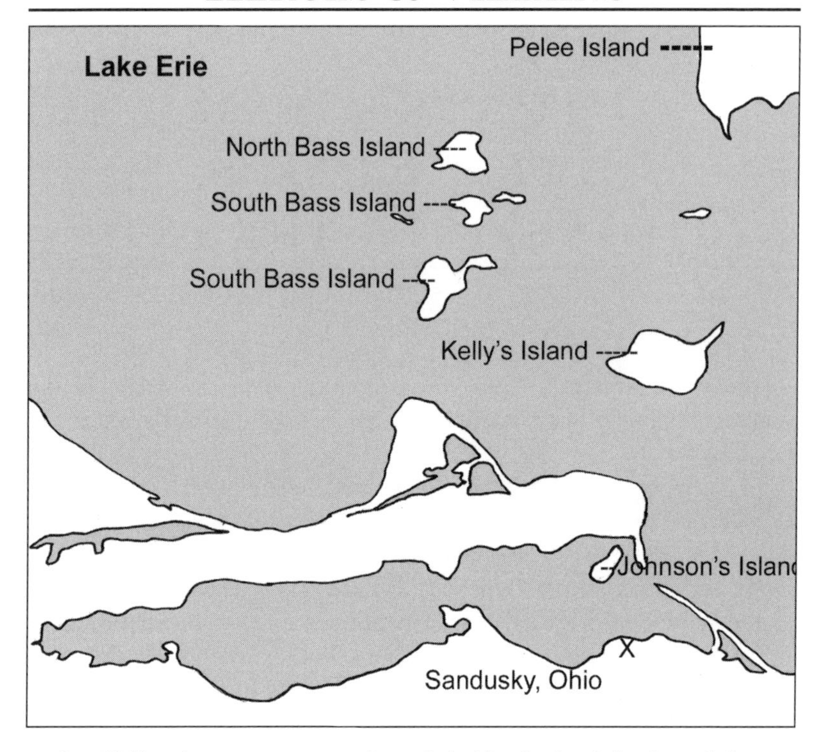

In 1861, the government leased half of the island and began construction of buildings to house and maintain prisoners and two forts to protect it from invasion. In April, 1862, the first 200 of the inmates arrived and over the next three years more than 15,000 Confederate enlisted men, officers, spies, Northern deserters and political prisoners were housed on the island.

In 1864, two Confederate conspirators, Charles Yates and John Beall, developed a plan to release the prisoners held at Johnson's Island. If their attack was successful, the daring invasion of a protected Union stronghold deep in the north would give the Confederacy a much needed boost in morale.

On September 19, 1864, Beall and several others boarded the sidewheel steamer *Philo Parsons* at Detroit, Michigan. The ship's normal route would take it down the Detroit River into Lake Erie for stops at North, Middle, and South Bass Islands and at Kelly's Island before reaching dock at Sandusky, Ohio. But after departing Detroit, John Beall persuaded the captain to make an unscheduled stop at the Canadian Port of Amherstburg, Ontario on the Detroit River.

About 30 passengers boarded at Amherstburg, all associates of John Beall.

The ship continued on it's course as scheduled with stops to discharge freight and passengers at the islands. After departing Kelly's Island, John Yates and his band of Confederates surprised the crew of the *Philo Parsons* and took command of the small steamer in an act of "piracy."

Mr. Frederick Hukill, a prominent Cincinnati businessman boarded the *Philo Parsons* at Kelly's Island that September afternoon. Mr. Hukill and his traveling companion, Mr. Skinner, had made their way to the upper deck just forward of the pilothouse. They found two chairs and settled down. He later related the following to the New York Times:

> "We lighted cigars and preparing to enjoy the little cruise to the utmost when something made me look up. For a moment I couldn't imagine what was happening for a man was holding a pistol to the head of the pilot. Then another man came forward with an explanation.
>
> "Stand back! Stand back!" he cried. Well, we didn't offer to move for in his hand he held one of the longest, prettiest revolvers it was ever my fortune to see, and he had us covered to a nicety.
>
> "What does this mean?" asked Mr. Skinner. "It means, responded John Beall, that we have captured this boat in the name of the Southern Confederation and you are my prisoners."
>
> Before anything else was said something happened to impress us with the stern reality of our situation. The engineer of the boat who must have attempted to resist his captors tried to reach the upper deck. There was a report of a revolver as he climbed the stairs. As he reached he top stair he fell almost at our feet and lay there with a bullet through his shoulder."

Captain John Beall's plan was to steam into Sandusky Harbor, attack the prison and release the prisoners. But, as the band of conspirators neared the harbor they saw the American gun ship *U.S.S. Michigan* anchored near Johnson's Island.

The *U.S.S. Michigan* was an imposing vessel. She was the United States Navy's first iron hulled warship. During the Civil War she was armed with a 30 pound Parrott cannon, five 20 pounders, six 24 pounders, and two 12 pound howitzers. The *U.S.S. Michigan's* primary

The U.S.S. Michigan *patrolled the Great Lakes during the Civil War. As John Beal approached Johnson's Island he found the ship anchored nearby. From the United States Coast Guard Historian's office.*

function was to protect the United States from any Confederate invasion which might come through the Great lakes from Canada.

John Yates Beall changed his plan of attack. The *Philo Parsons* would steam into Sandusky Harbor as it always did and surprise the sailors aboard the *U.S.S. Michigan*, overpower them and capture the ship. Then the released prisoners could use the gunship to escape.

To accomplish this, the *Philo Parsons* would need to take on more wood to fuel its boilers. Beall ordered the ship to return to Middle Bass Island for fuel.

En route to Middle Bass, the *Philo Parsons* came upon the small steamer *Island Queen*. The conspirators boarded and took command of the ship and its passengers. They would use the *Island Queen* in their attack on Johnson's Island as well.

The passengers and crew of the *Island Queen* were let off on Middle Bass Island as the fuel bunkers of the *Philo Parsons* were filled.

The *Philo Parsons* departed Middle Bass Island with the *Island Queen* in tow. The *Island queen* was towed a short distance when there was an apparent change of plans. The sea cocks of the *Island Queen* were opened and the ship was set adrift to sink near Chicanolee Reef.

The *Philo Parsons* steamed towards Sandusky Harbor to attack the *U.S.S. Michigan* and free the thousands of Southern prisoners held at Johnson's Island when suddenly Captain Beall ordered the attack halted.

It is still to this day not known why he stopped the assault, perhaps someone from shore signaled him to retreat, or possibly he did not receive an expected signal from shore, but he changed course and headed the *Philo Parsons* towards Canada to make their escape.

Had the Southern conspirators continued on with their mission, they would have been in for a surprise. The Union forces had been tipped off about the plan to free the Johnson's Islands prisoners and had laid a trap to thwart any attempts.

Captain John Beall was later hunted down and captured. He was tried and convicted of piracy and spying and sentenced to death. He was hung in March of 1865.

Dan Seavey, Pirate Of The Great Lakes

There was a man on the Great Lakes who was arrested and charged with piracy. He is the Great Lake's most famous pirate. The man was a hard drinking, hard living man named Dan Seavey.

Dan Seavey was born in Portland, Maine in 1865 and at age thirteen ran away from home to see the world from the deck of a ship. He enjoyed the life until he was eighteen when he joined the United States Navy.

After an undistinguished naval career, he obtained a job with the Federal Office of Indian Affairs, operating in Wisconsin and Oklahoma. His responsibilities included preventing the smuggling of whiskey onto Indian Reservations. Although there were not any laws prohibiting the consuming of alcohol in the United States, there were strict laws prohibiting the Native Americans, forced onto reservations, from drinking any alcoholic beverage.

"Roaring Dan," as he was called, was a tall husky man whose mere presence in a room commanded attention. Seavey was often involved in barroom fights. His large stature and quick temper lead to a reputation as a man to challenge if you wanted to enhance your reputation or to avoid if you favored your teeth. He lived on the edge of the criminal element.

Among other occupations Dan Seavey was engaged in was a stint as a gold miner in Alaska. The adventure proved to be just that, an adventure; he returned to the Great Lakes penniless.

Big Dan Seavey couldn't be kept down and he soon obtained the money to purchase the 40-foot schooner, *Wanderer*. On the *Wanderer*, Dan delivered cargos throughout Lake Michigan. He gained the reputation as a skillful sailor, one that would venture out in any weather conditions and he was usually the last to leave the lake just as winter set in. But, he also had the reputation as a sailor who would steal anything not bolted down.

In 1908, Dan Seavey was involved in an incident that sealed his place in American history; the arrest of Roaring Dan Seavey for Piracy on the Great Lakes.

On June 17, 1908, Captain R.J. McCormick master of the schooner *Nellie Johnson* returned to the dock to find his ship was gone. He summoned the Grand Haven, Michigan, police to report the theft of his schooner but the police did not believe the captain and did little to search for the vessel.

The police did nothing due to the condition of the captain... he was extremely drunk. He had spent the better part of the day in a local saloon and when he called the police they just couldn't believe the ramblings of a drunken sailor that his ship had been stolen.

In the following days, Captain McCormick was able to convince the authorities that the crime had indeed been committed and that the *Nellie Johnson* had been absconded by pirates!

The theft of the schooner was reported to the United States Revenue Cutter Service, the forerunner to the present day Coast Guard, and the cutter *Tuscarora* was dispatched to search for the pirated ship.

Dan Seavey and two of his men had taken the schooner *Nellie Johnson* with her cargo of cedar posts from Grand Haven, sailed it to Chicago and attempted to sell the cargo.

Roaring Dan and his "band of pirates" were not able to secure a buyer for the load of cedar post aboard the *Nellie Johnson* so they departed Chicago and sailed north along the Michigan coast to their home port of Frankfort, Michigan.

The *Nellie Johnson* was spotted by the crew of the Frankfort Life-Saving Station who notified the cutter *Tuscarora* that the ship they were in search of was in Frankfort. The pirated ship was moored upriver near Captain Seavey's home

Roaring Dan Seavey, was forewarned of the Revenue Cutter steaming towards Frankfort and slipped out in the middle of the night on his schooner *Wanderer*. His freedom was not long lived. About seven miles south of Frankfort, Dan Seavey was arrested.

Shackled in chains, Roaring Dan Seavey, the loud obnoxious scoundrel who had a reputation of stealing anything not bolted down, was bound over to stand trial on the charge of Piracy on the Great Lakes, a charge that if convicted carried a penalty of death.

Seavey's defense was that Captain McCormick, master of the *Nellie Johnson* had given the ship and its cargo to him in payment for a gambling debt. The prosecutors weren't buying his defense making a case for Piracy on the Lakes.

The case was followed in newspapers around the lakes and across the country. Reporters exaggerated facts or thought up new ones to sell the story, such as that Seavey had out drank Captain McCormick and his crew before he stole the ship, or that Seavey had tied Captain McCormick up in chains and thrown him off the ship to a watery Lake Michigan grave.

The much publicized trial turned out to be not as dramatic as anticipated. The prosecution's star witness, Captain R.J. McCormick, master of the *Nellie Johnson,* did not show up at the trial to testify against Dan Seavey and the case of Great Lakes Piracy was quietly dismissed.

Dan Seavey returned to upper Michigan more boisterous than ever. He again turned to sailing the *Wanderer* throughout Lake Michigan and transporting cargos sometimes with questionable bills of lading.

After the *Wanderer* was destroyed by a fire, Dan Seavey purchased a forty-foot motor vessel. Seavey took advantage of the speed of his new boat and was able to deliver more cargo.

His near death experience of almost being hung as a pirate didn't seem to settle the big man down. He was said to have become involved in many situations where he lived on the edge of the law and crossing the line on more than one occasion.

It is reported that Dan and his men illegally hunted deer in great numbers through the Upper Peninsula and shipped the venison off to lower port cities for sale. He is also accused of stealing deer meat, venison, from meat packers and selling it.

With Dan's reputation as a man of dubious character and his familiarity with the criminal element around the lakes, he was the perfect person to get into bootlegging when America entered the era of Prohibition.

While never caught, he was reputed to be a prime player in the transportation of Canadian liquor on Lake Michigan to the thirsty port cities of Chicago and Milwaukee.

Dan Seavey was reported to have operated a gambling ship on the Great Lakes, and he has been accused of keeping a vessel anchored off shore of Michigan's Upper Peninsula and Northern Wisconsin, stocked with prostitutes to provide entertainment for the isolated lumbermen and miners.

The life of Dan Seavey might have been exaggerated over the years but what is known to be factual about Roaring Dan is enough to make him one of the more colorful characters in Great Lakes history.

The Canadian Pirate William Johnson

William "Bill" Johnson, born in Canada in 1782, was a Canadian who detested the British and their domination over his homeland. During the War of 1812, the war between the United States and Great Britain, Mr. Johnson aligned himself with the Americans to fight the English.

When the war broke out he was living in Canada, but being sympathetic to the American cause, William Johnson moved across Lake Ontario to the United States where he was employed with the secret service and given a permit to capture all British property he might find on Lake Ontario and the St. Lawrence. William "Bill" Johnson was a "Privateer."

In small light boats of shallow draft, William Johnson and his men stalked the opposition throughout the St. Lawrence in the Thousand Islands area. Their knowledge of the area gave them an advantage.

He and his men were successful in disrupting the British in anyway they could. They intercepted British supply boats on the St. Lawrence and redirected the stores to American Troops. They snuck into towns and robbed British Sympathizers and destroyed any property that might be used against them.

On one occasion, he attempted to blow up a large new British ship before it was launched.

Johnson and his men quietly rowed into an enemy harbor with a rudimentary torpedo to blow up the HMS St. Lawrence. They entered the harbor in the dark of night but at morning's light they found that the big new ship and the British fleet had already departed.

Since they operated primarily in the shallow Thousand Islands, Johnson and his men chose shallow draft boats. The boats, known as a gig, were open boats propelled by 12 long oars for speed in open water or 12 paddles used in narrow channels. With these vessels, they could sneak up on their prey.

William Johnson at one time attacked a small group of British boats and took possession of all of its belongings. Among other bounty taken from the British ship Bill Johnson found bags of mail. As they read through the mail bags they found a letter from the Governor Prevost at Montreal to the Lieutenant Governor at Toronto. The letter spelled out information that was vital to the American military.

At the conclusion of the war in 1815, William Johnson settled back into legal occupations in the Thousand Islands region until the Canadian rebellion of 1837 also known as the Patriot War.

At this time, Canada was still under the flag of Great Britain and a growing band of Canadian citizens wanted to be free of the domination.

The disgruntled Canadians attempted to throw out the British government by force and enlisted Americans sympathetic to their cause to join their ranks.

Bill Johnson had hated the British and their rule over Canada for decades and when asked for his assistance, he saw this as an opportunity to once again fight for Canadian independence.

Based on his experience during the war of 1812, Bill Johnson was commissioned as the "Commander and Chief of the Naval Forces and Flotillas of the Patriot Services of Upper Canada."

The incident that sealed William "Bill" Johnson's place in history and dubbed him "Pirate Bill Johnson," the pirate of the Thousand Islands, was the raid on the steamer *Sir Robert Peel*.

The ship, a wood burning side-wheeler, was making its way up the St. Lawrence river with eighty passengers and an unknown number of crew. The ship being an early steam vessel burned wood rather than coal and required a steady supply of hardwood.

The *Sir Robert Peel* stopped at Wells Island in the Thousand Islands area of the St. Lawrence River to replenish its supply of cordwood. Historians of today question if it was wise for the ship to have selected Wells Island because of its close proximity to settlements which were known to harbor Canadian Patriots.

Sometime during the night the passengers and crew aboard the *Sir Robert Peel* were rudely awakened by a band of men with blackened faces and wearing Indian feathers in their hair. The band robbed the passengers and demanded they leave the ship. The crew and passengers, most dressed in their nightshirts and nightgowns, followed orders and allowed the ship to be pirated without a shot fired.

The marauders then went room by room plundering the ship for valuables. It had been reported that some wealthy passengers were carrying several thousand English Pounds and other valuables.

The pirates, content they had taken everything of value, untied the ship and allowed it to drift downstream. The passengers huddled together on the small island watched as the pirates set fire to the *Sir Robert Peel*. The British ship burned into the night until the last of the flames met the water's surface and the charred hull slipped below the water.

The "Pirate of the Thousand Islands" had struck! They had slipped in during the dark of night as pirates so often did and plundered one of the Kings ships, an act that secured a place in history for "Pirate Bill Johnson," Canada's only real pirate.

The Prohibition Era: 1920-1933

On January 29, 1919, the Eighteenth Amendment to the Constitution of the United States was ratified prohibiting the, manufacture, sale, transportation and possession of intoxicating beverages. One year later, on January 29, 1920, national prohibition became the law of the land.

Drinking of spirits was not new to this continent. Europeans who first established colonies in North America brought with them cases of beer and other liquors. In fact they usually brought more beer than water.

The practice of drinking whiskey, rum and beer was not unusual. They drank these liquids because there wasn't much of a choice; juice was unavailable or very seasonal, milk required refrigeration and water was often contaminated. The liquors and beer distilled and brewed with easily obtained raw materials was sometimes the best or the only choice our forefathers had to quench their thirst.

In early American history, drinking of alcohol was not looked at with disdain but public drunkenness was. The first laws written in 1619 did not prohibit drinking rather they were written for those who became drunk. And it wasn't until 1697 that a law was written to prohibit the sale of intoxicants on Sundays.

Throughout the 1700 and 1800s, various temperance movements promoted, primarily through churches, voluntary prohibition of the use of intoxicants. The movement made some progress in the fight against "Demon Rum." One tool they used to get their message out was music. One of the favorite temperance songs of the time was "The Temperance Army."

Now the Temp`rance army's marching,
With the Christian's armor on;
Love our motto, Christian Captain,
Prohibition is our song!

Chorus
Yes, the Temp`rance army's marching,
And will march forevermore,
And our triumph shall be sounded,
Round the world from shore to shore.
Marching on, marching on forevermore,
And our triumph shall be sounded,
Round the world from shore to shore.

Chorus

Now the Temp`rance army's marching,
Firm and steady in our tread;
See! The Mother they are leading,
Marching boldly at the head.

Chorus

Yes, the Temp`rance army's marching,
Wives and Sisters in the throng;
Shouting, "Total Prohibition,"
As we bravely march along.

In 1851, Maine was the first to vote in statewide prohibition of intoxicants. By 1855, Michigan, Vermont, New Hampshire, Delaware, Indiana, Iowa, Minnesota, Connecticut, Rhode Island, Massachusetts, and New York had joined the list of "Dry" states.

While the individual state legislatures voted to make their states dry, most did little or nothing to enforce it, while other states determined the law to be unconstitutional and repealed it.

The onset of the Civil War ended the prohibition of alcoholic beverages in the United States. During the Civil War drinking by soldiers was accepted, often the commanding officer issued a small amount of liquor to his troops to raise morale and the use of alcohol for medicinal uses was a standard practice during the war years.

In the decades following the war, the country had enough problems rebuilding and unifying the north and south, much less worrying about the populace drinking alcoholic beverages. The preachers kept warning their congregations about the evils of Demon Rum, but attempts to prohibit the consumption of intoxicants on a national scale were temporally forgotten.

By the 1880s, a new movement emerged. The Woman's Suffrage Movement took up the fight for several social reforms; the use of tobacco, closing of theaters, labor issues, woman's rights and the closing of saloons.

The woman's movement saw the saloon as the preverbal "Den of Inequity". Only men, painted ladies that worked as bar girls and prostitutes were allowed in saloons. Saloons served liquor and beer and promoted gambling and other vices, which would damn a man's soul to hell.

Ministers of all denominations preached sermons filled with fire and brimstone opposing the evils of the saloon.

By the turn of the century, the battle to save man from himself and his lust for alcohol and all evils that it brings, had again gathered momentum. Several states again passed legislation closing saloons or prohibiting the use of alcohol.

Some of the biggest proponents of the war on alcohol use were the industrial giants. Henry Ford was staunchly against his employees drinking. He determined with his assembly line method of production that the employees needed to work fast and precise. He deemed workers who drank were potentially dangerous to themselves and their co-workers.

Henry Ford established the Ford Sociological Department in his company. Employees from the department visited the homes of workers to evaluate the employees' lifestyles. If they were found to be heavy drinkers or of questionable morals, the employee might not receive raises or be fired.

Henry Ford was not alone in his thoughts of employees drinking, Ramson Olds' Reo Car Company hired detectives to keep an eye on their employees. If they drank, smoked or were active in social organizations that promoted the use of alcohol, they could lose their jobs.

One of the Reo factory supervisors was so dead set against use of beer and liquor that he refused to sell delivery trucks to breweries.

Michigan once again passed legislation making it a dry state. The law took effect at the stroke of midnight on May 1, 1918. It spelled the end of drinking in the state, at least the end of legal drinking.

In the days leading up to the May 1 deadline, the bars and saloons of Michigan were filled with patrons, some in drunken merriment and others mourning for the loss of an old friend… "John Barleycorn."

The law which closed Michigan's drinking establishments, made the possession of intoxicants illegal within the state boundaries, but it turned several thousand of the state's normally law abiding citizens into criminals. Michigan's large immigrant population who came to this country for freedom were suddenly criminals for following customs from their homeland; the German immigrants whose culture included a frothy stein of beer, the Italian population who considered wine part of their life and the Irish who were raised on whiskey became criminals for living as their forefathers had for centuries.

The Michigan government's biggest problem with enforcing prohibition was that alcohol was not prohibited in neighboring Canada or Ohio. For the thirsty Detroiters it was just a short drive down Dixie Highway to Toledo, Ohio, or even a shorter trip across the Detroit River to Canada.

Both Canada and Ohio were willing and able to provide all the booze Michigan could drink.

The distilleries of Canada and the breweries of Ohio didn't even have to deliver their products; there was no shortage of Michiganders eager to transport the booze. They were known as "Bootleggers."

The term bootleggers came from the cowboys who would shove bottles of whiskey into their tall boots to smuggle it into the Indian Reservations of Oklahoma where alcohol was prohibited.

Most bootleggers in Michigan at this time weren't hardened gangsters, they weren't organized crime connected, they were common men and woman just trying to make a few extra bucks.

PROHIBITION IN MICHIGAN

Goods, both legal and illegally, have been smuggled across the borders of the United States and Canada for centuries and the practice continues today. But smuggling on the Great Lakes reached it's pinnacle during the era of prohibition.

Since Michigan was dry and Canada and Ohio were not, there were huge amounts of money to be made. The governments of Ohio and Canada did little or nothing to stop their residents from selling beer and whiskey to the thirsty Michiganders.

Cases of illegal whiskey stacked on the deck of a rumrunners ship. Schooners like this one sailed the Great Lakes meeting smaller boats which bought the booze and delivered to waiting trucks on shore. From the United States Coast Guard Historian's office.

Michigan citizens would make midnight boat trips to the Hiram Walker Distillery south of Windsor, Ontario, take a ferry from Port Huron, Michigan, to Sarnia, Ontario, row a boat across the St. Marys River to Sault Saint Marie, Ontario, or drive down the Dixie Highway to Toledo, Ohio, for a few bottles for their own use or for resale.

There wasn't a train from out of state that stopped at a Michigan station that the conductor and engineers didn't have a couple of bottles stashed away. The same was true of the commercial boats that tied up in Michigan ports. The sailors were either making a couple extra bucks by selling a few bottles or bringing in a couple bottles for their own use.

There was not any large scale bootlegging operation, no organized crime, just folks taking advantage of a law they felt was unjust. Enjoying a drink with family and friends was legal for decades then suddenly on May 1, 1918, it was a crime to even have alcohol in your possession. It was a law that most opposed.

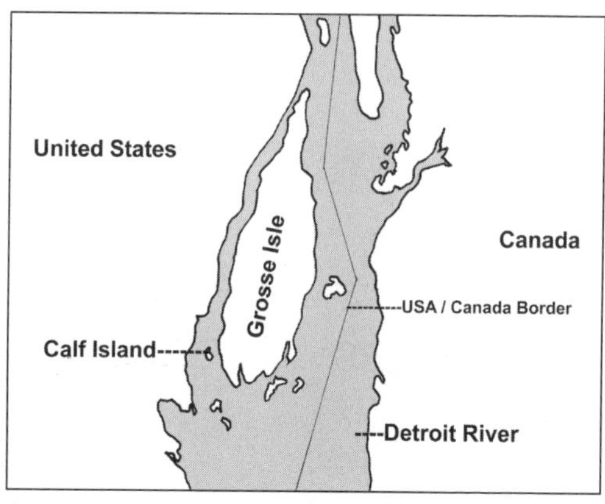

Michigan's Damon Act made the possession of intoxicating beverages illegal and the Wiley Act made the manufacture and sale illegal and shut down saloons.

The Damon Act made the possession of liquor illegal but it also gave the authorities the right to search individuals and their cars and homes without first obtaining a search warrant.

Police regularly pulled over cars on the Dixie Highway returning from Ohio, searched them and arrested the occupants if contraband goods were found. Homes were entered and searched based on rumor and individuals were taken off streetcars and searched because they looked suspicious. Many law abiding citizens of Michigan were unjustly detained.

GIN POSSESSION IN STATE LEGAL

Prohibition of Use of Liquor Worthless, Says High Tribunal, Upholding Dingeman's Ruling in Marxhausen Case.

SEARCH AND SEIZURE LAW INVALID; HOME IS A MAN'S CASTLE

Wiley Act, Banning Saloon, Is Good, But Holding Up of Autos for Inspection Is Utterly Illegal, Decision Asserts.

The summer home of August Marxhauson on Calf Island, located in the Trenton Channel of the Detroit River near Grosse Ile, was searched by authorities based on a rumor. Illegal intoxicants were confiscated and Mr. Marxhauson was arrested.

The Detroit newspaper millionaire vowed to fight his arrest and the confiscation of his property. His lawyers based their defense on the old English adage; "A man's home is his castle." They claimed authorities did not have the right to search Mr. Marxhauson's home without first obtaining a search warrant and by doing so they violated his rights.

LIQUOR FLOWS UNRESISTED INTO DETROIT

'Flivvers' Chase Each Other in Endless Procession.

74 CARS IN HOUR CROSS BORDER

Runners Declare That Condition is Too Good to Last.

Kinnane Confers With the State Officials.

MAY ASK AID OF CONSTABULARY

Dry Advocates Rush to Lansing to Get New Law.

After a well-publicized trial, Judge Dingeman from his Wayne County court ruled that the clause in the Damon Act permitting to searches without warrants to be illegal.

The ruling was appealed and finally landed on the bench of the Michigan Supreme court. The Supreme Court upheld Judge Dingeman's decision making the Damon Act Illegal!

The legality of the Wiley Act was upheld so it was still illegal for Michiganders to manufacture and sell alcoholic beverages, but the dissolution of the Damon Act made it legal for Michiganders to possess alcohol.

The decision of the Supreme Court had far reaching implications for the legal community in Michigan. Six men serving time in Jackson Prison for a second offense of the Damon Act were immediately released. More than one hundred persons held in county jails were released and the charges against more than 150 alleged violators waiting their day in court were dismissed.

The court ruled that the 4,500 men and women who had already served time or paid fines had no redress against the state.

August Marxhauson was awarded the goods confiscated from his summer home. The sheriff's department returned his 6,015 bottles of beer, 38 cases of wine, and 18 cases of miscellaneous alcoholic beverages.

The declaration of the Supreme Court opened the floodgates to thousands of Michiganders driving towards Toledo, Ohio, or Ontario, Canada, to purchase alcoholic beverages.

Although the new found ability for Michiganders to drink was not to last long. The 18th amendment creating national prohibition wasn't far off.

THE EIGHTEENTH AMENDMENT

The United States Constitution
Amendment XVIII

Section 1:
After one year from the ratification of this article the manufacture, sale, or transportation of intoxicating liquors within, the importation thereof into, or the exportation thereof from the United States and all territory subject to the jurisdiction thereof for beverage purposes is hereby prohibited.

Section 2:
The Congress and the several States shall have concurrent power to enforce this article by appropriate legislation.

Section 3:
This article shall be inoperative unless it shall have been ratified as an amendment to the Constitution by the legislatures of the several States, as provided in the Constitution, within seven years from the date of the submission hereof to the States by the Congress.

The eighteenth amendment to the Constitution of the United States was proposed to Congress on December 18, 1917, and was declared on January 29, 1919, to have been approved by 36 of the 48 States. The dates of ratification were: Mississippi, January 8, 1918; Virginia, January 11, 1918, Kentucky, January 14, 1918, North Dakota, January 25, 1918, South Carolina, January 29, 1918, Maryland, February 13, 1918, Montana, February 19, 1918, Texas, March 4, 1918, Delaware, March 18, 1918, South Dakota, March 20, 1918, Massachusetts, April 2, 1918,

Arizona, May 24, 1918, Georgia, June 26, 1918, Louisiana, August 3, 1918, Florida, December 3, 1918, Michigan, January 2, 1919, Ohio, January 7, 1919, Oklahoma, January 7, 1919, Idaho, January 8, 1919, Maine, January 8, 1919, West Virginia, January 9, 1919, California, January 13, 1919, Tennessee, January 13, 1919, Washington, January 13, 1919, Arkansas, January 14, 1919, Kansas, January 14, 1919, Alabama, January 15, 1919, Colorado, January 15, 1919, Iowa, January 15, 1919, New Hampshire, January 15, 1919, Oregon, January 15, 1919, Nebraska, January 16, 1919, North Carolina, January 16, 1919, Utah, January 16, 1919. On January 16, 1919 Wyoming became the final state necessary for the ratification of the amendment.

The amendment was later approved by Minnesota on January 17, 1919, Wisconsin, January 17, 1919, New Mexico, January 20, 1919, Nevada, January 21, 1919, New York, January 29, 1919, Vermont, January 29, 1919, Pennsylvania, February 25, 1919, Connecticut, May 6, 1919, and New Jersey, March 9, 1922.

Rhode Island was the only state in the union that did not approve the amendment.

The amendment eliminated all drinking of alcoholic beverages but it did not make provision for enforcement. Enforcement for the authorities was disorganized and the funding was lacking so it quickly became a nightmare.

The federal organizations charged with the enforcement of the law included the United States Customs Service, the United States Department of Treasury, a special prohibition unit, the Bureau of Internal Revenue and the United States Coast Guard. The law enforcement agencies on the state and local level were state police, county sheriff departments and local police departments.

One problem the authorities had in the battle with the smuggling of illegal booze was that enforcement organizations were competing for funding resulting in a lack of cooperation and organization.

Another problem, which hampered the success of prohibition, was the fact that Canada was so close and that Canada did not have laws making the manufacture and sale of intoxicants illegal.

BOOTLEGGERS AND RUMRUNNERS

The states bordering the Great Lakes with a close proximity to the distilleries of Canada became a major thoroughfare for the transportation for quality Canadian liquor.

Cities, towns and villages on the Canadian side of Lake Ontario lie less than a forty mile boat ride from western New York. Lake Erie is all that separates the Canadian shoreline from parts of New York, Pennsylvania and Ohio. Lake St. Clair, while not a Great Lake, borders both Canada and Michigan as does the entire approximate 245 mile length of Lake Huron.

To complicate matters more are the rivers that border both countries. The Detroit River runs 30 miles from Lake Erie to Lake St. Clair and is less than a half of a mile wide at its narrowest point. The St. Clair River runs north of Lake St. Clair for over 35 miles to the open water of Lake Huron, and the narrow St. Marys River at Sault Ste. Marie also posed an inviting avenue for bootleggers

Some Canadians took advantage of the economic principle of supply and demand. If there is a demand for a product then someone will supply the product.

Hiram Walker purchased a large tract of land south of Windsor, Ontario along the Detroit River where he built a distillery to produce Canadian Club whiskey.

When the Volstead Act became law in the United States, Hiram Walker and Sons Distillery was in a prime location to cash in. They could not sell liquor in the States but they did sell it by the case or barrel to boats arriving at the company docks along the Detroit River. There was even a rumor that there was a pipeline from the Hiram Walker distillery under the river to the States. The rumor probably was started because there was a pipeline that carried mash from the distillery to the Hiram Walker farm.

Ontario approved a law in 1919 that made it legal for citizens of Ontario to possess liquor and beer for home use, but all bars and saloons were closed down. The Canadian prohibition laws also differed from that of the United States in that they did not legislate against the manufacture of alcoholic beverages but the breweries and distilleries could not sell their products in Canada.

This resulted in Canadian liquor being sold to Michigan bootleggers who smuggled it back in to Canada.

The new Canadian law prohibited the sale of liquor and beer to the Unites States but the bootleggers found an easy way to skirt the issue. The captain of the boat picking up the booze needed to produce documentation that the cargo being purchased was destined for a country where there were not any laws prohibiting it. Many boatloads of fine Canadian whiskey were picked up at Canadian docks to be taken to

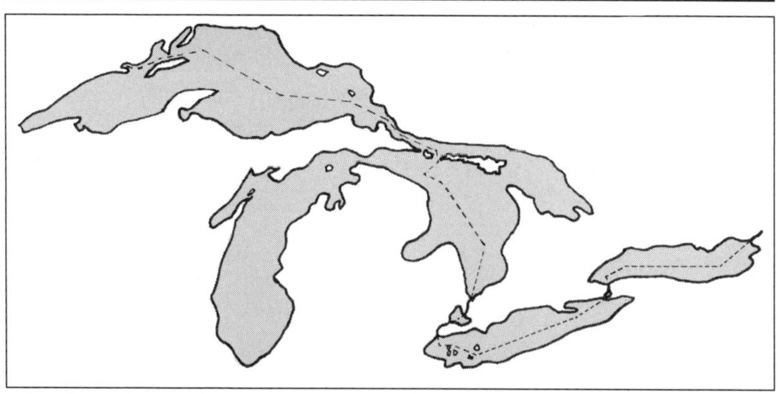

The dash line indicates the international boarder between the United States and Canada. The length of the border was the main obstacle in enforcing America's prohibition laws.

Mexico or Cuba. Yet in a matter of hours the boat had delivered its cargo to Cuba or Mexico and was back for another load.

When national prohibition was first enacted, the federal government determined the largest assault by bootleggers would come along the Atlantic coast and concentrated its efforts there.

The illegal intoxicants flooded into the east coast of the United States from Canada at the north and sailed up from the Bahamas, Cuba, and the Caribbean from the south. But they under estimated the potential of the Great Lakes.

While rum running was prolific in the cities that shared a river with Canadian cities, it was prevalent on all of the Great Lakes. None of the lakes were without their bootleggers considering Lakes Ontario, Erie, Huron and Superior share a boarder with Canada. Only Lake Michigan lies entirely within the geographical boundaries of the United States.

The majority of the liquor smuggled into the United States was brought into the country at populated areas along the Canadian/United States boarder. It is said that more booze entered the States across the Detroit River and Lake St. Clair than anywhere else including the Atlantic coast.

But the Great Lakes has over a thousand miles bordering Canada, and thousands of uninhabited and desolate coves, bays and beaches where a ship loaded with a cargo of contraband could enter and easily discharge its cargo. Once off the boat, the cargo would be loaded aboard waiting trucks and transported to the thirsty public throughout the United States.

Several fishermen enjoying some illegal beer during the heyday of prohibition era. From the collection of Oscar Ingles.

On Lake Huron, just 18 miles north of the busy St. Clair River lies the village of Lexington, Michigan. Today Lexington is a popular tourist town with a State of Michigan Marina, a private marina, fine restaurants, interesting shops and friendly people, but during the prohibition era the town was a typical lake port. A wood dock extended out into the lake used by small steamers to take on and discharge passengers and cargo. It was also where the town's fleet of fishing boats docked.

Only 8 miles across lower Lake Huron from Canada, Lexington became a prime location for smuggling illegal liquor into Michigan. Citizens of the area who disagreed with the ban on booze or who simply wanted to cash in on the large sums of money available in the bootlegging trade, engaged in the activity.

In 1928 a small fishing boat, under the cover of night, tied up to the Lexington dock. The men onboard quietly went about their business, trying not to attract any attention, for the cargo they had brought to the Lexington dock was over three thousand bottles of beer.

The men off loaded hundreds of cases of beer, placing them in stacks on the dock. They worked quickly and quietly but, unknown to them, government men were watching the entire operation. As the government officers approached, the bootleggers noticed and threw off the lines, started the engine and sped away leaving 2,400 bottles of beer on the dock.

The government officials couldn't leave the contraband just on the dock but they couldn't stay around and guard it either so they began tossing the cases of beer off the dock. The cases sank to the bottom, out of sight and out of mind. When they finished the government officials returned to Port Huron intent on bringing back more men to remove the illegal drink from the lake.

The operation did not go unnoticed by the citizens of Lexington. Several men changed into bathing suits and began diving off the dock to retrieve the bounty that lay on the bottom.

Treading water they took a deep breath into their lungs and dove down only to return to the surface with armfuls of bottles of fine Canadian beer and a smile on their face.

As the pile of bottles on the dock grew, so did the crowd. Word spread quickly about the booze boat, the government officials and the retrieval of the beer. It wasn't too long before the dock was filled with people opening up and consuming the bottles of beer... chilled to the temperature of Lake Huron.

About ten miles up the Lake Huron shore lies the small village of Port Sanilac. Within the protected harbor are private and state marinas and a fleet of recreational sailboats, but during the years of prohibition fishing boats were the primary vessel at the dock. In the area there was a man who kept a still to produce a low quality liquor he would sell to locals.

A photograph of ships sheltering at the Harbor of Refuge at Harbor Beach, Michigan. From the Tony Lang Collection.

But he also used his knowledge of the lake to make several trips a month across the 35 miles to Bayfield, Ontario. He would purchase several cases of fine Canadian whiskey and beer and return to Port Sanilac. The cargo would be transferred to a waiting truck that transported the quality booze to the clubs and restaurants of Detroit.

To avoid being recognized on his many trips across the lake, he would repaint the boat a different color every couple of trips. It was said the boat had paint a quarter of an inch thick.

At Harbor Beach just 25 miles north along the coast, on August 22, 1928, there was a storm ravaging the lake. Many ships chose to avoid the wind and waves and put into the Harbor of Refuge at Harbor Beach. One ship which entered the harbor and approached the breakwall to tie up was the steamship *India*. As the ship eased its way to the breakwall, an 18 year-old deckhand, August

Maxwell jumped from the deck of the ship to the concrete surface of the breakwall. The deckhand slipped and fell backwards off the breakwall into the harbor.

The crew of the *India* and crews from ships nearby lowered their yawls and threw cork vests into the water but the young man had disappeared.

The lighthouse keeper at the Harbor Beach lighthouse, observing the ships at the wall was watching the rescue attempts and put a call to the Coast Guard Station.

Captain Davidson and a crew of the Harbor Beach Station responded to the breakwall in their motorized lifeboat. They maneuvered around the ships moored at the wall looking under the turn of their hulls and between the ships for the missing sailor, eventually finding the body of the young man.

Captain Davidson climbed aboard the *India* to get information as to how the accident occurred and about the unfortunate young man. While he was on the deck of the *India*, Captain Davidson noticed a fishing tug tied to the wall. It looked as though it might be taking on water for it was sitting very low.

Captain Davidson and his men motored over to the boat, The Bird from Michigan City, Indiana. Calls to the boat went unanswered so the captain and one of his men climbed aboard the fishing tug.

They looked about the craft for signs of the boat taking on water and their crew but only found one sleeping man and over 1,000 cases of Canadian beer! Captain Davidson reported there were cases in every free space.

They arrested the sleeping man and searched for the rest of the crew who apparently abandoned the boat when the Coast Guard was called to search for the deckhand of the *India*.

The fishing tug was moved from the breakwall to the coal dock where police guards were posted to protect the boat and its cargo until Prohibition Officers from Detroit arrived.

It was said that the fishing tug was so packed with cases of beer that one more case could not have been wedged in. The more than 1,000 cases of beer off loaded onto the coal dock was estimated to be valued at over $24,000.00.

Trucks were enlisted to assist in hauling the beer from the coal dock to an area near the beach. The cases filled with bottles of beer were stacked, oil poured over them and set on fire.

The crowd that gathered at the beach to witness the destruction of the illegal cargo rivaled the crowds at the Fourth of July celebration. Some onlookers cheered the capture of the illegal beer while others mourned its loss.

Over one thousand cases of Canadian beer were removed from the Bird and taken to the beach to be destroyed, but according to town folklore, it was far fewer cases that made it to the fire. Mysteriously several cases disappeared between the coal dock and the beach.

About 12 miles north of the Harbor of Refuge at Harbor Beach is Whiskey Harbor.

Whiskey Harbor was not populated and quite away off the main road. It was a perfect spot for bootleggers to bring ashore illegal booze. The contraband was brought in on small boats, typically owned by locals who made the two-hour trip across the lake to Goderich, Ontario. Yet in the later years of prohibition, the area was also used by mobsters who, in small boats, met north and southbound steamers carrying the illegal cargo. The small vessels met steamers in the dark of night then transported the cargo back to Whiskey Harbor. The cargo was quickly loaded on waiting trucks that drove the goods to Northern Detroit.

Ironically, despite its historic roll in the bootlegging business Whiskey Harbor was not named because of the rumrunners, it was so named because years earlier a passing steamer with a cargo of lumber sank near there and some bottles of whiskey washed ashore. The locals named the small cove Whiskey Harbor.

There was an ingenious man on Wisconsin's Door Peninsula in Lake Michigan who found a way to legally sell booze during prohibition.

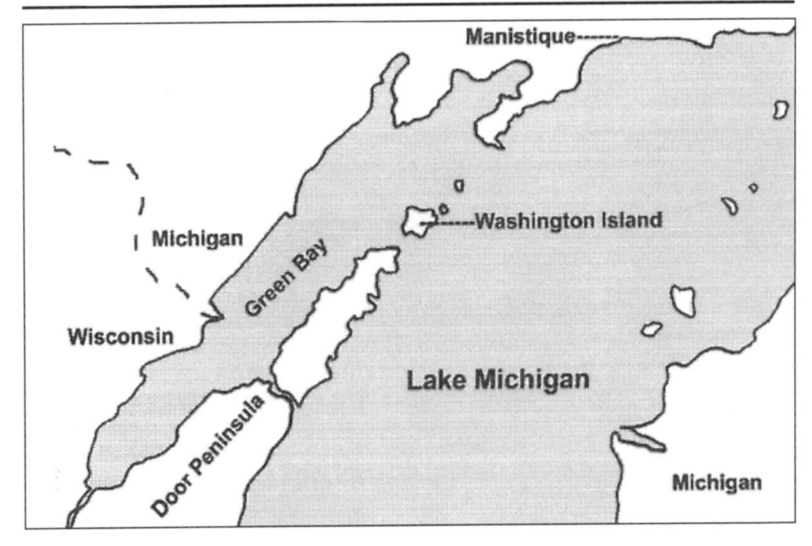

Tom Nelson was the owner of Nelson's Hall on Washington Island. Prior to prohibition Nelson's Hall was a favorite place for citizens and visitors to the Island to enjoy a Saturday night dance, drink a beer or down a shot or two. But Tom Nelson's business took a sharp decline after the 18th Amendment became law.

Mr. Nelson was told he should sell bitters which was legal during prohibition. He did some research and found that bitters were introduced in 1824 as a medical remedy for stomach ailments. Bitters contain up to 45% alcohol but because of their medicinal purpose they were not classified as liquor.

Since it was a medicine, only a doctor or pharmacist could dispense bitters. Tom Nelson, being an enterprising business man, applied to become a pharmacist.

Throughout the prohibition era, Tom Nelson sold bitters to the poor unfortunate citizens and visitors of Washington Island who suffered from ailments of the stomach.

Bootlegging was not the domain of just men; there were women involved as well. In Detroit, women smuggled booze and sometimes drugs across the border. They would take the ferry across to Canada and return with their contraband hidden from the eyes of the Custom Agents. The agents were less apt to check a woman. But a bottle or two of booze might be under a baby's blanket, wrapped up in a diaper, or stuffed in secret pockets in her petticoats.

An individual who showed ingenuity went by the name "Muskrat" LaFramboise. Muskrat was a man of questionable character who eked out a life before prohibition fishing and trapping along the south end of the Detroit River.

With his knowledge of the river and not being averse to breaking the law, "Muskrat" LaFramboise was an ideal person to become involved in the bootlegging business.

As the demand for illegal beverages became greater, Muskrat's cargos became larger, his trips to Canada more frequent and his reputation grew. Those wanting to buy booze knew his bootlegging ability, but his reputation also alerted the authorities to his operation.

"Muskrat" LaFramboise used all of his talent and knowledge of the waters to smuggle in his illegal contraband. But running at night, hiding in little known coves, and using fast boats wasn't always enough to avoid the police so he developed a new technique.

The ingenious Muskrat drilled holes in the bottom of his boat and installed valves. When a Coast Guard, sheriff or police boat snuck up on him or when he determined that capture was inevitable, he opened the valves, flooded the boat until it sank and he swam away. After the

Muskrat LaFramboise opened valves in the bottom of his boat to sink it and its incriminating evidence. But, some bootleggers chose to burn their boat and the booze rather than be caught. From the United States Historian's office.

authorities departed, Muskrat went to the site of the sunken boat, dove down and recovered his cargo. Muskrat was indeed enterprising, yet a man living on the edge of society and the law.

When the warmth of summer gave way to winter the demand for alcohol didn't dry up so bootleggers had to change their mode of operation to deal with the conditions. They waited for the rivers and lakes to freeze over and form an ice bridge to Canada.

Winter rum running was not easy. Bringing the booze across the frozen rivers and lakes was dangerous. The bootlegger, pulling a sleigh or driving a truck across the ice, faced the possibility of falling through the ice, getting lost in a blizzard, frostbite, being hijacked by other crooks and Prohibition officers.

Some bootleggers walked across the narrow rivers pulling sleds piled high with cases of whiskey or other intoxicants. The more daring rumrunners bought old jalopies and drove them across the ice to Canada to take on a load of liquor. The danger of the car or truck breaking through the ice was ever present. In fact, the bootleggers took the doors off the vehicles just in case they had to make a quick jump from a car breaking through the ice.

One bootlegger told the story of lying flat on his stomach on the ice waiting for the police to leave. Someone, probably a rival bootlegger, had tipped off the police that a load was coming across the river. The bootlegger saw them and laid down while the police shined their flashlights across the ice. In the cold and blowing snow the police left after several hours, then the bootlegger continued on his way.

The Prohibition officers, Border Patrol agents, police, Sheriff's Deputies and State Police were all engaged in the battle against the rumrunner. It was said catching a bootlegger during prohibition was like an old cat trying to catch a young mouse.

The Border Patrol, founded in 1925, was a new agency. They tried to chase down bootleggers with old junk cars when the bootleggers drove new cars with supercharged engines and heavy duty suspensions. And the weapons they were given were surplus Colt revolvers from the First World War.

Deputy Duncan McKenzie of Bad Axe, Michigan, was photographed pouring gallons of homemade moonshine down the sewer. From the collection of the Bad Axe Historical Society, Bad Axe, Michigan.

The mid 1920s were a tough time. Jobs were hard to find and only paid 20 to 30 cents per hour. Many honest people were lured into running rum because a bootlegger could fill a car with whiskey in Canada for a few hundred dollars and sell the load for $1,000.00

To further hamper the efforts of the law enforcement officers, half of the country disagreed with the law and many would do nothing to help the police. Those living along the rivers bordering Canada and the Unites States may not have been actively involved with bootlegging but many helped the bootlegger out. Boathouses of citizens were used for receiving illegal cargo, citizens left a light on as a signal that the police were near, or they looked the other way when a boat pulled up to their dock and a car pulled up in their driveway.

Many stories of rumrunners and their exploits became public after the 18th Amendment was repealed. One story related the tale of an enterprising bootlegger.

The police confiscated a keg of Canadian whiskey and moved it to a dock on the St. Clair River to be later disposed of by dumping it into the river. An armed guard was posted to keep the curious and thirsty away. When the Prohibition officers later arrived, they opened the bung to pour out the keg's contents and found it empty. They examined the keg and found a hole had been drilled from below the dock and the keg's contents were drained into containers in a boat below the dock.

Another story of bootlegging occurred in Port Huron, Michigan. The bootlegger was delivering a keg of beer to a second floor speakeasy. Just as he got to the top of the stairway, the police burst in through the ground floor door. The bootlegger was startled and let go of the keg. The keg bounced down the stairs, the police scattered, the keg rolled out the door, and smashed into the window of the department store across the street, and the bootlegger got away.

The rumrunners found that if they transferred the bottles of whiskey from the wood cases into straw filled burlap bags they could pack more bottles into a boat. In addition if the bootlegger detected the presence of police they could throw the sacks overboard and they would sink to the bottom to be retrieved later. If the bottles were left in their original wood cases they floated. Some youngsters were known to dive down for a bottle or two before the bootlegger returned or the kids were hired by the rumrunner to retrieve the cargo.

There are reports from S.C.U.B.A. divers to this day finding cars and trucks on the bottom of rivers bordering America and Canada. There are even rumors of the vehicles being packed with cases of fine Canadian whiskey, which it is said still tastes great.

The procedure the Prohibition Service followed was that after the contraband was seized and cataloged the booze or beer was poured out. After a raid some towns had a river of beer flowing down the street. They often used the same area to break the confiscated bottles of beer or liquor. Piles of broken glass could be found in most cities where bootlegging was prevalent.

Lake Erie has approximately an 213-mile long border separating fine Canadian whiskey from the thirsty public of the States. There were many citizens from the shore towns of Toledo, Cleveland, Sandusky, Erie and others cities, towns and villages headed across the lake to purchase some of what the 18th Amendment made illegal in Ohio.

Buffalo, New York in the 1920s was an industrialized city with a diverse ethnic population. Europeans immigrating to the United States sailed across the Atlantic Ocean landing in New York. Many remained there but many more sought their livelihood in the Midwest portion of the new nation. They left New York City and traveled by railroad or canal boat to Buffalo where they could board another ship to sail to the cities, mines and lumber camps around the Great Lakes.

There were opportunities for work in Buffalo and some immigrants stayed on there. At the turn of the century, the largest ethnic group in Buffalo were of German descent. Breweries making beer from old German recipes sprang up throughout the city to support the saloons and bars the German population frequented. When the 18th Amendment was ratified making the manufacture, sale and consumption of alcohol illegal, the German residents of Buffalo were in an uproar.

Beer was a part of their heritage, part of their culture. They had been raised on beer, how could the government take it away from them?

The result was that Buffalo was known as a "Wet" town during the "Dry" years of prohibition. The enforcement of prohibition laws was largely ignored. During the 1920s, Buffalo's Mayor Francis X. Schwab was very concerned when some of his constituents were killed or blinded by drinking bad moonshine. Speaking with a thick German accent he denounced the prohibition laws and on two different occasions tried to get 18th Amendment repealed or at least modified. He reasoned that if the manufacture of liquor was controlled and regulated by the government, then the quality of the liquor would be kept to a higher standard. His appeal fell on deaf ears and there were no changes to the law.

U.S. District Attorney William J. Donovan was upset over the lack of enforcement and disregard for the law by the citizens and the mayor of Buffalo and was determined to make changes. In 1922 Donovan made a statement by having Mayor Schwab arrested and charged with manufacturing and selling beer.

Donovan prepared for a lengthily and hotly contested trial and openly said that if the mayor was found guilty then he should be given the most severe punishment possible. He wanted to send the message to the citizens of Buffalo that lawlessness would no longer be tolerated. A highly publicized trial was anticipated and newsmen from around the country were ready to cover the story

The popular mayor, much to Donovan's dismay, pled guilty, paid a five hundred dollar fine and the whole matter was dismissed.

U.S. District Attorney William J. Donovan was furious over the result of the Mayor Schwab affair and vowed to end the blatant disregard of the law.

FIGHTS RUM RUNNER, TAKES STOLEN YACHT

Bay City Manufacturer Regains by Force Craft Held for Liquor Smuggling.

OBTAINED UNDER A CHARTER

Owner Has Yacht Moved From Dock Before Law Can Prevent It.

The backroom saloons the common man frequented were plentiful and popular but Buffalo's elite had their own establishments: the County Club of Buffalo and the Saturn Club. The registry of both contained the names of most of Buffalo's elite. Lawyers, bankers, giants of industry, presidents of shipping companies, and even judges were members. They met at the clubs to transact business, socialize, eat the finest of meals and smoke hand rolled cigars. They also went to the clubs to consume some of the finest beer and liquor available.

On August 29, 1923, Federal agents under the direction of U.S. District Attorney William J. Donovan, raided the Saturn Club and the Country Club of Buffalo. The people arrested read like the Who's Who of Buffalo high society.

Those arrested could easily afford to pay the fine and most were back at the club the following day. Donovan's raid only succeeded in making him the most hated man in Buffalo. Booze and beer still flowed in Buffalo.

The case of the stolen yacht in Bay City, Michigan, demonstrates the lengths a rumrunner will go to get his product into the United States. Mr. William Sovereign, the president of Bay City's Aladdin Company, chartered his seventy-two foot yacht to a well-dressed Detroit man who arrived at the dock in a luxury touring car. The deal was that the man wanted to charter the yacht for two weeks so his family could try it out.

Saturday night dances were a major source of entertainment in the 1920s. Dance halls similar to this could be found in any city, town or village.

Then if he liked it he would purchase it. As a security deposit the man left his expensive car.

After two weeks, the man called telling Mr. Sovereign the boat needed some hull work and it was in a dry dock in Detroit. He also informed him that he had changed his mind and was not going to buy the boat and wanted his car back. Mr. Sovereign grew suspicious and went to check the dry docks in Detroit, but he could not find his yacht. Further investigation showed that the yacht was recorded when it passed through the Welland Canal at Port Colborne, Ontario, on Lake Erie. Canadian authorities were contacted to be on the lookout for the seventy-two foot yacht. The ship was found in Toronto, Ontario. When Government lawyers told Mr. Sovereign that it could take up to a year to get his boat back, he took matters into his own hands.

He traveled to Toronto, located his yacht, and gathered some local thugs to assist him. The band of men approached the ship just as rumrunners were beginning to load a cargo of 500 cases of whiskey aboard the ship. When Mr. Sovereign and his gang of thugs appeared the rumrunners ran. The Bay City Manufacturer had his yacht back.

MICHIGAN'S WHISKEY REBELLION

In the Michigan Upper Peninsula village of Iron River, prohibition became a problem for its residents. Iron ore was discovered in the area in 1880 and the village grew with the development of the mines. The demand for laborers brought an influx of immigrants escaping oppression in their homelands or seeking a better life in the new world. They flocked to the Upper Peninsula for the plentiful jobs in mining and lumbering.

Iron River prospered with its diverse and multi-cultural population. John Scalcucci and his brothers, Joe and Steve immigrated to Iron River making the village their home. They worked hard and by the 1920s they owned a packing company, a boarding house, a claim on land for mining and a grocery store. They operated the store on the first floor, lived on the second and made wine in the basement.

Just months after the 18th Amendment took effect, the Michigan State Constabulary (Forerunner of the Michigan State Police) raided the Scalcucci store and two other homes of Iron River residents of Italian decent and seized barrels of red wine commonly referred to as "Dago Red."

A local state attorney, Martin McDonough, was called by the constabulary asking where they could store the evidence but McDonough was outraged that the Michigan Constabulary raided the homes without first obtaining a warrant. He determined the raids were illegal and ordered the wine returned to the homeowners.

The Michigan State Constabulary was angry that a State's Attorney would not support them in their efforts to rid Iron River of the brothers who made barrels of wine in direct opposition to the 18th Amendment.

A few days later State Constables again raided the Italian - Americans but this time they had a federal prohibition agent with them. The constables and agent confiscated the same barrels of "Dago Red' they were ordered to return just days before.

Mr. McDonough showed up at Scalcucci's store with local police and protested the raids and demanded the wine to be returned to the local residents. He claimed the only thing going on in Iron River that was illegal was the search and seizure of private property without a warrant.

Leo Grove, Supervising Prohibition Agent for Michigan's Upper Peninsula claimed that when McDonough arrived at the Scalcucci's store, he threatened Grove and arrested him and drove him out of town.

Agent Grove went directly to Chicago to report to the Prohibition Supervising Director of the Central District, Major A.V. Dalrmple. He told of the insolence in the small village of Iron River in Michigan's Upper Peninsula and that they were revolting against the 18th Amendment. He also told Major Dalrmple the States Attorney was leading the rebellion against the Federal government.

In February of 1920, prohibition was only in its second month of existence. Throughout the nation there were many people who were openly defiant of the law and gallons of illegal intoxicants flooded into the States but this was the first occasion where a States Attorney, a government official appointed to uphold the laws of the State of Michigan and the United States, publicly opposed the law of the land.

Headlines of newspapers all around the nation told of how the small mining village of Iron River, Michigan, was in an open revolt against the United States government's laws prohibiting alcoholic beverages. The papers dubbed it "Michigan's Whiskey Rebellion."

The *Chicago Tribune*, February 23, 1920, headline read, "Whiskey Rebellion; U.S. Defied in Michigan."

Major Dalrmple told reporters that he would go to Iron River with as many prohibition agents and Michigan Constabulary as necessary and put down this blatant disregard of the laws of the nation. He vowed he would arrest the State's Attorney McDonough and the deputies who defied the authority of Agent Groves.

Word of an army of law enforcement personnel preparing an invasion reached the residents of Iron River. Reporters from newspapers across the nation flooded the small town in preparation of the first bloody battle of the Whiskey Rebellion. But when the reporters arrived, they found white flags, white towels, white shirts, white sheets, anything white the citizens of Iron River could find flying from the houses in surrender.

The white flags flew as many barrels of wine, beer and whiskey were secretly being moved to hiding places in the woods, caves and mines.

State's Attorney Martin McDonough told reporters that the citizens of Iron River are not rebels intent on defying the United States, they were just simple people who wanted to maintain their way of life.

He explained to the assembled reporters that he was merely doing his job by protecting the rights of the residents of Michigan. He determined the raids on the homes in Iron River were performed illegally because the Michigan State Constabulary and the Prohibition agents did not first obtain search warrants as is required by law.

McDonough explained that when he arrived at the Scalcucci Brother's store he found Prohibition Agent Leo Grove with barrels of wine he had confiscated loaded on a sleigh. When asked for identification to prove he was indeed a Prohibition agent all he could produce was a soiled and tattered letter of his appointment; he did not carry the identification card as all agents were required to and when Grove began to move the sleigh, McDonough who questioned the validly of the appointment letter, placed him under arrest for transporting liquor. Shortly afterward Grove was released.

He also explained to the reporters that the Scalcucci brothers had taken a delivery of grapes to be sold in their store. When the grapes did not sell they followed the recommendation printed in an Internal Revenue pamphlet and made wine. The man said the wine was for their own personal use, it was not to be sold.

McDonough pointed out that Section 29 of the Volstead Act exempts the domestic manufacture of non-intoxicating cider and fruit juices from the prohibition law so long as it is neither sold or delivered for consumption.

McDonough said he was not in rebellion with the nation and the prohibition laws, he was in rebellion with the disregard of state and federal officials to follow search and seizure laws.

Major A.V. Dalrmple arrived by train in Iron River with sixteen armed men and hundreds of rounds of ammunition. They came prepared to fight the rebels who stood in defiance of the Prohibition Laws.

Newspapers exaggerated the facts of the story, they had reported that: The Village of Iron River was filled with foreigners intent in fighting the prohibition laws to the last drop of blood. If the prohibition agents came to the Upper Peninsula of Michigan, they would be met by armed men determined to fight for their right to drink alcoholic beverages.

Major Dalrmple, furious that insurrection was occurring in his district requested a warrant for the arrest of States Attorney McDonough. The United States Commissioner in Marquette, Michigan, agreed that warrants should have been obtained prior to the raids and refused to issue a warrant on McDonough.

Dalrmple boasted that he would use his own judgment as to the legality of raiding dwellings suspected of keeping illegal intoxicants and in a display of force Major Dalrmple and his men raided the home of the local priest. They confiscated barrels of wine that were in a locked storage room in his basement. The agents took the barrels out to the road, broke them open, allowing the wine to pour into the ditch.

McDonough confronted the Major and told him that if he enters another dwelling without a warrant that he and his men would be arrested. McDonough would not stand for anyone violating the rights of his constituents.

Newsmen telegraphed exaggerated stories back to their home cities telling of the armed standoff taking place and that violence and bloodshed was sure to follow.

In Washington D.C., the news stories were being closely followed. Fearing that the agents were not on solid legal ground, the Commissioner told all agents that from now on if they did not have the cooperation of local authorities they must first obtain a warrant raiding private homes. They also warned their agents, including Major Dalrmple, about exceeding their authority.

Politicians in Washington D. C. and Michigan were outraged at the thought of armed warfare breaking out between law enforcement officers and residents of a small Michigan village. Word was received in Iron River that the Assistant Chief of Prohibition Enforcement from Washington D. C., the United States District Attorney from Grand Rapids, Michigan, and the Assistant Attorney General of Michigan were all en route to Iron River.

Major A. V. Dalrmple suddenly declared that he must leave immediately for urgent business in Washington D.C. He and his men abruptly left the small village of Iron River, while the reporters wrote articles blasting the whole episode as a charade.

They wrote that the armed show down between Dalrmple and the lawbreakers of Iron River led by a State's Attorney was nothing more than a hoax perpetrated on the people of the nation.

Dalrmple left in defeat but the local State Attorney Martin McDonough was heralded as a hero. He bravely took on the federal government to protect the rights of the residents of his district. He received telegrams and letters from all across the nation congratulating him and offering him jobs.

On February 28, 1920, just two weeks after the first illegal raid on the Scalcucci's Brother's grocery store, the federal government declared that their investigation into the Iron River raids was finished and that no charges would be brought. The great "Whiskey Rebellion of Michigan" was over.

During prohibition one enterprising young man from Marquette, Michigan developed an ingenious way he could transport and sell the liquor he produced in his backwoods still.

Whenever he drove his car to the dance hall, he made sure to park in an out of the way location shielded as much as possible from view. He also made sure that when he parked one of the air valves on his tires was low.

This young man had filled the inner tube of his tires half to three quarters full with the booze. Once the inner tube was placed back into the tire he would over inflate the inner tube and put the tires back on his car.

Then when a customer wanted to purchase a pint of his whisky he would take a glass jar out to his car, open the valve and the pressurized air in the inner tube would push out a stream of his finest backwoods liquor.

BEN KERR - A CANADIAN RUMRUNNER

Ben Kerr was a man with the brains, the daring, the right background, and was in the right place to become a very successful rumrunner. Ben grew up as an outdoorsman. Young Ben learned many lessons from his father and uncle; they hunted, fished and he learned to read the lake and sky. These were skills that would be of great use for him in future years.

Ben was a rebellious child and quit school when he was just a teenager to work for a stonemason. Two years later he was fortunate to be selected as an apprentice plumber. But Ben had other ideas. He saved his money and bought a parcel of Lake Ontario waterfront property in Hamilton, Ontario. He borrowed money and built boathouses where the wealthy residents of Hamilton could leave their powerboats. His boat storage business was very successful until other Hamilton businessmen went into competition with him. Then to attract boaters he had to slash prices and he was barely making his payments.

This boat is similar to the style of Ben Kerr's Evelyn. *From the United States Coast Guard Historian's office.*

Ben was on the brink of bankruptcy, he needed funds to save his boathouse business and to keep a roof over his family's head. He turned to the quick money to be made by becoming a bootlegger. He had a boat and prohibition was being enacted just across the lake in the United States. He was perfectly placed to bootleg Canadian liquor and beer to the Americans.

Toronto, Ontario located on the northern shore of Lake Ontario was said to be ideally located for the rumrunner. Products of the breweries and distilleries of Canada were easily taken across the lake to the thirsty Americans. Bootleggers using fast powerboats took on loads of booze in small Canadian fishing villages or isolated coves and raced across the lake.

His talent and daring made Ben into one of Canada's best rumrunners. In the early 1920s he was operating two boats, one which took delivery from Colby's Distillery, transported it to his other, larger boat which Ben himself piloted across the lake to Rochester, New York, or other harbors along the lake's southern shore.

The notorious Canadian gangster Rocco Perri became aware of Ben's exploits. Rather than fight among themselves the two worked together. Kerr exported liquor to the States but also delivered it to Canadian ports for Perri's speakeasies.

Business was so good that Kerr had a local boat builder build him two new speedboats, the larger of the two was 35-feet in length, named Evelyn. The Evelyn made several trips across the lake per week making Ben Kerr one of the biggest suppliers of fine Canadian whiskey.

Kerr was making big money and he invested in another boat. In 1924 he took possession of a new boat, the *Martimas*. The *Martimas*, was capable of carrying 1,200 cases of beer or liquor, and had a wood hull covered in steel. The steel covering could break through ice which allowed Kerr to run across the lake even after most boats, including the Coast Guard, had been stored away for the winter.

Ben knew the American Coast Guard kept improving the vessels they were using to intercept the bootleggers crossing Lake Ontario. So Ben had to keep updating as well. His next boat, the *Pollywog*, was a custom built forty foot wood vessel with steel plating for traveling through winter ice. Equipped with two 180 horsepower engines, the *Pollywog* was a fast boat capable of outrunning hijackers and giving the Coast Guard a run for their money.

One day Ben Kerr and an accomplice loaded the *Pollywog* and made a late February run across Lake Ontario. The following day the pair did not return as anticipated. But since sometimes Ben returned to another port to take on cargo, no one panicked. But a week later when there was still no word from the two bootleggers, fears grew that they might have met with trouble on the lake.

Friends and relatives of the two men patrolled the shore hoping they might find the *Pollywog* trapped in ice. Private airplanes flew over the lake looking for any sign of the boat. For a while it was suspected that the two men might have made it to Main Duck Island, but an inspection of the island found the missing men were not there.

Almost a month later, parts identified as being from the *Pollywog* were found on the shore as spring melted the lake ice. Then the grizzly discovery of the remains of Ben Kerr and his accomplice were found nearby on shore.

Had Ben Kerr crossed a line and angered Rocco Perri or other gangsters who ran booze across the lake? Had another bootlegger attempted to highjack Ben's *Pollywog* and make off with its cargo of booze? It was never determined who killed Ben Kerr. The life of a bootlegger can be very profitable but it is down a dangerous road they travel.

A SHIP'S CARGO ENDS UP ON THE BEACH

In November of 1922, the steamer *City of Dresden* had locked through the Welland Canal from Lake Ontario into Lake Erie. The papers of the ship, which was of questionable sea worthiness, said it was bound for Mexico. It was common for bootleggers to buy a boatload of booze and say it was headed for some area other than the States. The old ship was actually bound for Detroit, Michigan with a cargo of 1,000 cases and 500 kegs of Corby's Whiskey.

As the *City of Dresden* made its way across northern Lake Erie, a westerly storm developed and as so often happens on the shallowest of the Great Lakes, the seas grew quickly and large. Captain McQueen directed his vessel behind the lee of Long Point, Ontario, in the calmer waters of Gravelly Bay.

As the ship lie at anchor, the westerly blow swung around until it was blowing from the northeast. Long Point no longer offered the protection from the storm, rather the change in direction now threatened to blow the decrepit ship, loaded with thousands of pounds of whiskey, on the shore.

Long Point is a peninsula about 19 miles long that extends out into Lake Erie from the Canadian shore. The point has long offered a refuge from the lake's storms, but it also was a nightmare to many ships navigating the shallow reefs surrounding the point.

The storm whipped the lake into a maelstrom. Waves grew to eight to ten feet and assaulted the *City of Dresden* as she lie at anchor.

Captain McQueen knew if he didn't move the *City of Dresden*, the anchor would drag and the ship would end up nothing more than a pile of kindling on the beaches on the shore of Long Point.

The old engines had seen better days and could not develop enough steam to power the *City of Dresden* away from its anchorage. The ship had to be lightened. Captain McQueen ordered the crew to push part of the deck cargo of whiskey kegs into the lake, hoping the engines could then power the ship to safer waters.

As the waves beat on the ship, some of the seams of the wooden ship parted allowing lake water to pour into her cargo hold. The ship's steam pumps worked in an effort to empty the bilge.

The *City of Dresden* was able to maneuver around the southern point of Long Point where she was afforded some protection but her anchors could not find a secure bite in the sandy bottom and the waves continued to beat on the ship.

A lone Life-Saving Service surfman patrols the shore watching for signs of ships in distress. From the United States Coast Guard Historian's office.

The *City of Dresden* drifted helplessly towards the shore when she grounded hard on the sand bar off shore and began to break apart.

Captain McQueen sounded the distress signal to notify those onshore they needed help. The Long Point citizens who heard the steam whistle of the *City of Dresden* screaming for help pulled on their slickers and made their way to the beach to see the steamer on the sand bar just a few hundred yards off shore.

The crew of the *City of Dresden* lowered a yawl in an attempt to get off the ship but the driving snow and high winds ripped the small boat from the hands of the crew and was blown away. The ship started to break apart as the waves continued to smash down on it.

As the growing crowds on the beach watched, the crew lowered another small boat. All of the crew boarded the small boat but the waves overturned it, casting the crew into the lake. All of the crew swam back to the small boat except the son of the captain who was swept away and not seen again until his lifeless body was found several miles away.

The oars of the small boat were washed away; the crew could only drift along at the mercy of the wind and waves. The boat drifted in the seas with the waves mounting the boat from behind threatening to overturn it while its occupants prayed to make it to the safely of shore.

Several people on shore saw the boat and threw them a rope, then risked their own lives by entering the raging lake to pull the survivors to shore.

Most of the crew had safely made it to land while the waves beat the *City of Dresden* into pieces. The crowds on shore turned their attention to the ship's cargo that began to wash up on shore.

A Canadian Life-Saving Station patrolman walking the beach on the east side of the point saw something drifting to shore and watched it until he realized it was a keg of whiskey. Then another and yet more kegs came ashore on the waves crashing in the shallows. The kegs the captain ordered thrown overboard to lighten the ship were washing up. The patrolman dragged and rolled the kegs beyond the wave break and buried them in the sand. He and his friends would later dig them up.

As the ship was broken apart by the huge waves, cases, kegs and bottles of the cargo came ashore along the southern end of the western side of the peninsula. Word quickly spread that a cargo of whiskey was washing up on Long Point. Hundreds of people made their way by car, truck or horse drawn wagon to the southern end of the Point to get their share of the cargo.

People loaded the bounty on their vehicles, some buried the whiskey in the sand and others simply built a fire on the beach and opened a few bottles for a party. Bootlegging can be very profitable but in the case of the *City of Dresden* the residents of Long Point were the only ones who profited.

MOONSHINERS

Not all booze was smuggled into the United States from Canada, Europe or the Caribbean; a lot of illegal hooch was distilled in the States in homemade stills. The distilling of alcoholic beverages was prevalent in the United States long before the age of prohibition. To early Americans alcohol was not just a drink but they needed to distill liquor out of necessity. It was used as a disinfectant and as a medicine for almost any illness that could affect mankind.

The homemade liquor went by various names, Moonshine, Corn Liquor, White Lightening, Skull Cracker, Ruckus Juice, Rotgut, Mule Kick and Hillbilly Pop to name a few. By any name, it was dangerous to consume. Without government regulation, moonshiners were not held to any standards and consuming moonshine was sometimes fatal.

An example of a backwoods still. From the author's collection.

To distill moonshine, the raw materials needed were few and easily obtained. To make 36 gallons of corn liquor, the moonshiners needed 50 pounds of cornmeal, 200 pounds of sugar, 12 ounces of yeast, and 200 gallons of water.

Mix all of the ingredients together to form the mash. In a large metal container heat the mash to the point the liquid vaporizes, trap the vapor in a coiled tube cooled by cold water and the resulting liquid captured in a container is the moonshine. The process is not difficult, doing it in the backwoods, hiding from the "Revenuers", and getting the finished product to market was the hard part.

Some unscrupulous moonshiners wanting to speed up the fermentation process, included some additional ingredients. Relying on their rudimentary knowledge of chemistry, they knew by adding ingredients like lye, sulfuric acid, battery acid, or animal manure, the fermentation of the mash could be accelerated.

Moonshiners also cut their product (adding some other inexpensive liquid to increase the volume) by adding such liquids as industrial alcohol, turpentine, and paint thinner.

Not all moonshiners were guilty of adulterating their product. Many made it like their grandfathers before them and took great pride in the quality of their Corn Liquor.

But those who consumed bad booze met with a grueling agonizing death. They often went blind as the toxic chemicals used to cut the liquor crept into their brain. Severe stomach pains followed as the chemicals ate away at their intestines. They often suffered mercilessly until they died.

When bad booze found its way into the hands of the public, outraged citizens screamed for the authorities to crack down on the bootleggers who brought in the poison. The citizens demanded that those who were responsible be brought to justice. Even the police officers, prohibition agents and judges who were receiving payoffs from the criminals couldn't overlook the crime. They had to find who was responsible and bring them to justice.

Stills could be found everywhere. The stereotype of the backwoods moonshiner being from the hills of the southern states is just a stereotype. Moonshine was produced all across the nation and in Canada. Although most were located in rural areas close to larger cities, some booze was made in the cities. The phrase "Bathtub Gin" comes from booze made from mixing a concoction of chemicals in the bathtubs of homes and apartments. The quality was low and the alcohol content high.

Just outside Detroit, Revenue agents raided a Pontiac, Michigan, farmhouse that they suspected harbored a moonshine operation. When they arrived, a man ran from the house and escaped despite the hail of gun fire from the agents. Upon inspection they found parts of a still which had apparently been dismantled and moved. They did find the moonshiner's hiding place for his stash of illegal booze. The agents found a fifty-five gallon barrel of whiskey under a pile of manure. The moonshiner built a device to siphon the booze from the barrel to fill glass jars. The agents got the booze but the moonshiner and the still got away. The still was probably back in business before the agents left the farmhouse.

In the "Thumb" area of Michigan the Huron County Sheriff, John Hoffman and two deputies, Duncan McKenzie and Roy Hicks, entered a home of a suspected moonshine operation. When they entered the man of the house drew a large knife and waved it at the Sheriff, while Deputy Hicks walked into a room where the moonshine was stored. There he found the 240 pound "Mrs" standing with a broad axe held over her head and aimed at the deputy's head.

Deputy Hicks drew his pistol and fired at the woman's feet and she lowered the axe in surrender. Both the "Mr and Mrs" were arrested and three quarts of whiskey and two barrels of mash were confiscated.

In Minnesota, one county produced a moonshine that became known across the country for its quality. In many speakeasies it was asked for by name, "Minnesota 13."

At the end of World War I, much of the United States experienced an economic depression. Since Stearns County, Minnesota, relied on agriculture as its main economic source they were hard hit. The farmers sometimes were forced to sell their harvest for less than it took to grow it. Many Stearns County farmers turned to making moonshine just to put food on the table.

BAD RUM KILLS 27 IN 2 STATES

More Deaths Are Expected in Connecticut and Massachusetts.

Seven Arrests Are Made—Probe Traces Liquor to Hartford and New York.

There were several factors responsible for the quality of the liquor produced in Stearns County: the German heritage of the residents of the county, the moonshiner's desire to make a quality product, and the use of good quality ingredients.

When foreigners immigrated to the United States they tended to settle in areas where the climate was similar to that of their homeland. Thus many people of Germany settled in northern midwest agricultural areas such as Minnesota.

The German immigrants came with the knowledge to distill liquor and a work ethic to do it right. If they were going to produce illegal whiskey, they would produce the best quality whiskey they could. They would use the best ingredients and build their stills of quality materials.

Some unscrupulous moonshiners used galvanized metal tubs for fermenting their mash. The tubs were often soldered with lead that leached into the product, sometimes at levels that would poison the drinkers of the hooch made in the still. In Stearns County, the stills were made from stainless steel and copper.

To produce the quality moonshine, they took some time consuming steps often overlooked by most moonshiners. They took the time to age their product in charred oak barrels.

The last factor affecting the quality of liquor distilled in Stearns County was the corn used, "Minnesota 13."

The University of Minnesota, as early as 1888, began to develop a seed base that was hardy and could be grown in the short Minnesota growing season. A seed lot, number 13, showed promise and was further enhanced and marketed to Minnesota farmers under the name Minnesota 13.

It wasn't long before the news of the quality of moonshine being produced in rural Minnesota reached the gangsters of larger cities. Bootleggers made frequent trips to Stearns County from Minneapolis, Green Bay, and Milwaukee, Wisconsin. A distribution network was established to transport Minnesota 13 from Stearns County to the speakeasies of Chicago.

Isadore Blumenfeld, a.k.a. Kid Cann, the notorious Minnesota mobster, made the arrangements for the famed "Minnesota 13" moonshine to be picked up and delivered to Al Capone's clubs in Chicago.

The illegal but high quality liquor produced in Stearns County, Minnesota, became known as "Minnesota 13." From the small rural county, their moonshine was distributed as far away as San Francisco, California.

PROHIBITION REPEALED

The United States Constitution
Amendment XXI

Section 1:
The eighteenth article of amendment to the Constitution of the United States is hereby repealed.

Section 2:
The transportation or importation into any State, Territory, or possession of the United States for delivery or use therein of intoxicating liquors, in violation of the laws thereof, is hereby prohibited.

Section 3:
This article shall be inoperative unless it shall have been ratified as an amendment to the Constitution by conventions in the several States, as provided in the Constitution, within seven years from the date of the submission hereof to the States by the Congress.

Today

War Near in Europe! Hillman's Amazing Revelations Start Today

Turn to Page 4

Herald and Examiner

Chicago

METROPOLITAN EDITION

WEDNESDAY, DECEMBER 6, 1933

PRICE 3 CENTS

PROHIBITION ERA ENDED! LOOP CROWDS HAIL REPEAL

U.S. STUDIES 'FAVORITISM' IN HOME LOANS

Order Extradition of Touhy Gang for Factor Trial Here

NATIONAL CITY BANK FOR U.S. MONEY POLICY

Utah Records Final Vote; U.S. Proclamation Follows

U. S. Rushes New Liquor Control System.

QUOTAS FIXED

20 Million Gallons in Canada Ready for Entry.

President Signs Death Warrant of Prohibition

Proclamation Declares Dry Amendment Repealed; Calls for Temperance.

Groups of Girls Line up at Hotel Bars.

DINERS JOVIAL

No Boisterousness; Wild Driving Banned.

The twenty-first amendment to the Constitution was proposed to the several states by Congress on February, 20 1933, and was declared on the 5th day of December, 1933 to have been ratified by 36 of the 48 States.

The dates of ratification were: Michigan, April 10, 1933, Wisconsin, April 25, 1933, Rhode Island, May 8, 1933, Wyoming, May 25, 1933, New Jersey, June 1, 1933, Delaware, June 24, 1933, Indiana, June 26, 1933, Massachusetts, June 26, 1933, New York, June 27, 1933, Illinois, July 10, 1933, Iowa, July 10, 1933, Connecticut, July 11, 1933, New Hampshire, July 11, 1933, California, July 24, 1933, West Virginia, July 25, 1933, Arkansas, August 1, 1933, Oregon, August 7, 1933, Alabama, August 8, 1933, Tennessee, August 11, 1933, Missouri, August 29, 1933, Arizona, September 5, 1933, Nevada, September 5, 1933, Vermont, September 23, 1933, Colorado, September 26, 1933, Washington, October 3, 1933, Minnesota, October 10, 1933, Idaho, October 17, 1933, Maryland, October 18, 1933, Virginia, October 25, 1933, New Mexico, November 2, 1933, Florida, November 14, 1933, Texas, November 24, 1933, Kentucky, November 27, 1933, Ohio, December 5, 1933, Pennsylvania, December 5, 1933 and Utah, December 5, 1933.

The amendment was later ratified by Maine, on December 6, 1933, and by Montana, on August 6, 1934. The amendment was rejected by South Carolina, on December 4, 1933.

The approval of the 21st Amendment to the Constitution of the United States repealed the 18th amendment and ended 13 years of national prohibition.

Prior to prohibition, the prevailing thought was that if alcoholic beverages were not available, most of America's social ills would disappear.

During one his sermons after the passage of the 18th Amendment the Reverend Billy Sunday said:

> "The reign of tears is over. The slums will soon be a memory. We will turn our prisons into factories and our jails into storehouses and corncribs. Men will walk upright now, women will smile and children will laugh. Hell will forever be for rent."

However nothing could be further from the truth.

Consumption of liquor decreased in 1920, the first year of prohibition, yet studies show that the consumption quickly rose above pre-prohibition levels.

More Americans were drinking hard liquor as opposed to beer. It was easier to import whiskey, gin, rum or other liquors than beer. A case of beer took up just about the same amount of space on the rumrunner's boat as did a case of liquor, but the liquor commanded several more times the price. So mostly liquor was bootlegged into the States and the American drinking habits switched from the lower alcohol content of beer to the more potent liquor drinks.

Prior to prohibition, the manufacture of liquor was held to high standards, yet during prohibition the moonshine and bathtub gin were unregulated. Without government intervention, drinking illegal beverages was often a dangerous undertaking.

Many people were sickened or killed by unscrupulous bootleggers who adulterated quality imported liquors by cutting them with other chemicals to increase the volume and maximize profits. Also, quality control at backwoods stills was unheard of. Dead rats and other animals were often found floating in the stills and barrels of the moonshiner.

The Temperance Movement saw the saloons as dens of inequity filled with all sorts of temptations. The Movement succeeded in closing the legal saloons but turned the saloons into speakeasies and other secret drinking clubs. In fact in most major cities in the United States, there were more drinking establishments during prohibition than before it.

It was often said that it was harder to find a drink before prohibition than during it.

Prisons were not turned into factories as Reverend Sunday predicted. Just the opposite happened. More prisons were built to house the increase in the criminal element. By 1932, inmates in prison for violating

The police display a storage room of slot machines and still evidence from illegal gambling and moonshining. From the Bad Axe, Michigan Historical Society.

federal laws had increased 561%, homicides in the United States during the prohibition era increased 78% over pre-prohibition years, and violent crimes against persons and property were at an all time high during the prohibition years.

Another unexpected outcome of prohibition is that it turned normally law-abiding citizens into criminals. In the early years, citizens who disagreed with the 18th Amendment based on the fact that they considered a drink of wine with dinner, a stein of beer with friends or a bit of whiskey with their mates to be part of their cultural heritage. These Americans were forced into becoming criminals by driving or taking a boat ride to Canada to buy their alcohol or by visiting their local speakeasy for a drink.

To this day, many wealthy families in coastal towns owe their riches to ancestors who made nightly boat trips across a river or out onto a lake to meet a dark ship in Canadian waters, or maybe the ancestor drove a truck across the ice loaded down with cases of liquor or brought a case or two in on a train.

One of the worst results of the prohibition era was the development of organized crime. The huge profits that could be made smuggling booze into the United States drew in the penny anti thieves and street thugs. The more enterprising crooks united beyond the neighborhood gang stage and organized into crime syndicates that quickly took charge of all illegal activities: gambling, prostitution, drugs, extortion, murder and booze.

Villains Of The Great Lakes

Crime has always been a part of any civilization and the Great Lakes region is no exception. Petty thieves, murderers, and organized gangs have and still exist in the region. During the prohibition of the sale and consumption of alcohol, the immense profits that could be made by breaking the law and bringing booze into the United States was a temptation that few criminals could resist.

Many individuals became bootleggers to supplement their income or to get drinks for their own uses. Most of this criminal element remained small time bootleggers. But the huge profits of rum running attracted the hardened criminal who ruthlessly did whatever it took to reap the huge profits bootlegging offered.

Purple Gang

During the prohibition years, there were many individuals and several gangs that operated in the Detroit area. The close proximity of dry Detroit to the rapidly flowing liquor of Canada made the area ripe for rum running. Some of the Detroit area gangs were the Oakland Sugar House Gang, the Third Street Gang and the most infamous of the gangs, the Purple Gang.

The Purple Gang began as a group of young Russian Jewish immigrant boys who had moved to the industrial city of Detroit with their parents and found that living in the land of milk and honey was not all it was advertised to be. Jobs, if they could be found, were low paying and the housing they could afford was usually in run-down neighborhoods; a breeding ground for thieves.

The four Bernstein brothers, Abe, Joe, Ray, and Izzy were the recognized leaders of the gang who began as vandals and shoplifters, stealing from shopkeepers and strong-arming other youths, but their

Members of the notorious Purple Gang who violently ruled Detroit's prohibition era illegal activities with an iron fist. From the author's collection.

criminal activities expanded to more violent crimes, as the boys grew older. It was said that they got their name from a shopkeeper after the thugs stole from his store. He said the youths were rotten, they were purple like rotten meat.

Robbery, breaking and entering and extortion were soon crimes the gang excelled at but they would also strong-arm businessmen and rival gangs through the use of such methods as beatings, bombing, and murder.

When prohibition was enacted, the gang was quick to turn their attention to the money they could make selling liquor in the United States.

The Purple Gang found that there was more risk in smuggling the booze into the States than in selling it once it was in the States, so they made the distribution and sale their specialty. They purchased it from bootleggers, but more often they would hijack loads of liquor from individuals and other gangs. The Purples would not only steal the load of illegal alcohol but they would do it with Thompson machine guns blazing.

Because Detroit was so close to Canada, huge quantities of booze flowed across the river. It all wasn't consumed in Detroit, rather the booze was distributed all over the Midwest. Soon gangs from other cities looked at the profits that could be made by running rum across the Detroit River and they wanted a piece of the action. This led to bloody fights between the Purple Gang and other gangs.

By 1928, the Purple Gang had made millions of dollars in illegal activities but their own devices soon broke them up.

The *Detroit Free Press* headline of September 17, 1931 screamed of the murder of three gangsters. Like the St. Valentine's Day Massacre in Chicago, Detroit's criminals had brutally slaughtered three rival gang members.

Hymie Paul, Joe Sutker and Joe Lebowitz had last been seen alive at a club with Solly Levine and were said to be in high spirits when they left. The three men had been at odds with the Purple Gang for suspicion of hijacking the Purple's trucks of booze, but the men were happy because Solly Levine had brokered a peace agreement and a meeting with the Purples.

Ray Bernstein, Irving Milburg, Harry Keywell and Harry Fleisher, all members of the Purple Gang, greeted the four men at the Collingwood Manor Apartments, apartment 211.

Levine, Paul and Lebowitz sat on the couch with Joe Sutker on the arm. Cigars were passed out and drinks poured. The conversation was light and Levine and the others relaxed. Bernstein made an excuse to leave and went to the car, started the engine and revved the engine loudly.

The loud engine was the signal for Irving Milburg, Harry Keywell and Harry Fleisher. When they heard it, the Purples pulled out their guns and began firing.

When the smoke cleared, Hymie Paul, Joe Sutker and Joe Lebowitz lay in a large pool of coagulating blood, all three were still holding their

WEATHER

The Detroit Free Press

1831 ~ A CENTURY OF SERVICE ~ 1931

METROPOLITAN FINAL EDITION

THURSDAY, SEPTEMBER 17, 1931 — THREE CENTS

3 KILLED BY GANG: 5 JAILED

LEGION MEN TO PASS ON JOB CRISIS

O'Neil to Arrive Here from Conference Thursday

VARIOUS PLANS ARE SUGGESTED

Labor Insurance Urged by Swope

General Electric Head, Asking Reforms, Warns Manufacturers to Forestall Laws Menacing Business

NEW YORK, Sept. 16—(A.P.)—A co-operative management-employe plan aimed to eliminate unemployment, stabilize business, and remove the worker's fear of idleness and old age, was offered industry tonight by Gerard Swope, president of General Electric Co.

10,000 VIEW WOODWARD FIRE FIGHT

Blast at Grand River Sends Sheet of Flame in Street

BLOCK PERILED; THEATER BLAZES

M'COLL DUE TO GO NEXT

Removal from Jury Board Position Is Sought

SUES AIMEE'S LATEST MAN

MYRTLE ST. PIERRE

VICTIMS HERDED INTO APARTMENT, BODIES RIDDLED

Known to Police Here as Handbook Operators and Engaged in Rum Racket on Big Scale

Shot in Backs—Four Men Speed from Scene in Collingwood Manor at Rate of 70 Miles an Hour

Detroit had its version of Chicago's St. Valentine's Day massacre Wednesday afternoon, when three liquor racketeers, who had defied a rival gang's warning, were kidnaped in the thick of W. Grand Blvd. traffic, herded into an apartment hired for the killing, and slaughtered in cold blood.

Five alleged associates of the dead men were questioned by prosecutor's officers early Thursday morning in an effort to sift out of their devious underworld dealings a clue to the reason for the murders and to the identities of the killers. The men held for investigation are: Sol Levine, 30 years old, Fort Wayne Hotel; Otis Broder, 46, of 837 Owen Ave.; David Cohen, 32, of 470 Stimson Ave.; Louis Goldberg, 30, of 9325 Broadstreet Ave., and Roy St. Clair, of 674 Brainard St.

The triple murder occurred in Apartment 211 of Collingwood Manor, 1740 Collingwood Ave. The victims were identified by fingerprints in the Police Identification Bureau as Joseph (Nigger Joe) Lebowitz, alias Leibold, 31 years old, 660 Brainard St.; Joe (Izzy) Sutker, alias Sutton, 28, of 949 Reed Place, and Hymie Paul, 21, of 881 Merrick Ave. All had police records.

cigars. Solly Levine stood looking at thee bodies waiting for one of the mobsters to turn towards him and take aim, but rather they asked if he was ok.

A witness said she watched out her kitchen window disgusted with a car near the back of the apartments revving its engine and causing it to backfire several times. As she watched four men ran from the apartment, jumped into the car and the vehicle sped away screeching its tires on the pavement.

The killers had planned the assassination of the mobsters in advance. The .38 Colt pistols used had their serial number filed off and the mobsters had taken a bucket of green paint and placed it in the kitchen. After they had gunned down Sutker, Paul and Lebowitz, the murderers tossed the pistols into the bucket of paint so finger prints could not be detected.

The three killers and Solly Levine sped off. Solly was waiting for one of them to put a hole in his head. Later it came to light that Ray Bernstein had plans for Levine. The plan was to later kill Levine and plant evidence on him indicating he was the killer of Sutker, Paul and Lebowitz.

The three dead men were no choir boys caught in a gangland crossfire. Hymie Paul, Joe Sutker and Joe Lebowitz were brought to Detroit originally as enforcers for the Oakland Sugar House Gang to strong arm legitimate businessmen and collect from rumrunners. They soon tired of doing the work and not getting a large enough part of the profits. They fell out of favor with the other Detroit gangs partly because of their repeatedly hijacking rival loads of whiskey, encroaching on other gang's territory and killing at random. They were unpredictable, they held allegiance to no one. Their deaths were heralded by the Detroit underworld as a good thing, but not by the public.

The brutal murders outraged the citizens of Detroit and Prosecuting Attorney Harry S. Toy. An all out manhunt was mounted to find the men responsible for maliciously killing the three gangsters. Tips were phoned into the police, probably from rival gangs taking advantage of the opportunity to get rid of the Purples.

Their warring with other gangs resulted in bloody battles with bodies piling up until the public would no longer stand for it. The Detroit newspapers screamed for the violence to stop and for witnesses to step forward to break the hold the gang held on Detroit. Investigators were able to tie several crimes to Purple Gang members, sending many key members to prison and the effectiveness of the gang was soon diminished.

AL CAPONE

In Chicago, one of the country's great cities, one of America's most infamous criminals rose to power, Alphonse Gabriel Capone.

Born in 1899 in Brooklyn, New York, young Al Capone grew up in a world of crime. As a child he was involved in youth gangs and graduated to become involved with the crime families who wielded so much power there was little local police could do.

Al Capone was a tough guy. It is reputed Al had killed two gangsters, but not enough evidence was found to bring him up on charges. A bit later he severely beat a rival gang member, but before he was arrested Al was sent to Chicago and set up with a job working for "Papa" Johnny Torrio, the notorious Chicago gangster.

At that time, Johnny Torrio's South Side Italian Gang was locked in a bloody battle with the North Side Irish Gang run by Dion O'Banion for control of bootlegging and other rackets in the city. Torrio ordered the assassination of O'Banion, which set off an all out war for control of the city.

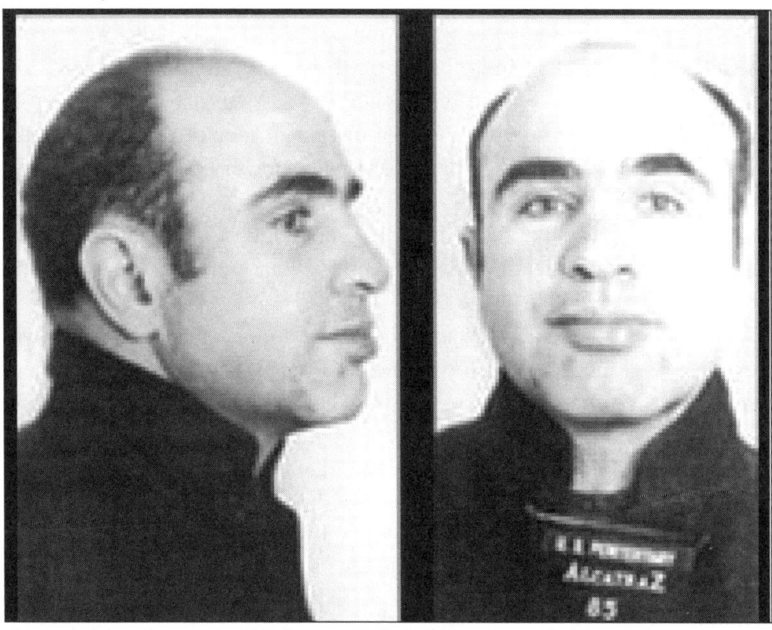

A mug shot of a young Alphonse Gabriel Capone. Al Capone's reputation grew to make him the most infamous criminal in the United States. His name is still synonymous with the Chicago underworld. From the F.B.I. Collection.

One day in November of 1924, three men walked into the flower shop owned by O'Banion. He greeted them with a smile and they greeted him with pistols. Dion O'Banion's death was just the beginning of the war between the gangs which would last for over four years and result in forty murders. Forty that are known as there are probably many more that went unnoticed by the police.

Mr. Torrio was impressed with Al's intelligence, brawn and fearlessness working in the organization. He promoted Al to supervise the gang's bootlegging business, a position ranking number 2 in the gang's hierarchy.

In retaliation for the murder of O'Banion, North Side gang members ambushed and nearly killed Johnny Torrio. Torrio survived the attack but decided to retire and move out of Chicago. It was Al Capone who assumed the head of the gang. Under his direction the organization's saloons, speakeasies, nightclubs, bookie joints, gambling and prostitution houses, racetracks, breweries and distilleries thrived. But the war with the North Siders was also kept hot with several more murders by both sides.

Al Capone handled any opposition to his iron fist rule in a very definitive way. He had them killed on the street outside their home or they disappeared never to be seen again. It was said that Al Capone was responsible for ordering the deaths of dozens of men during his reign as Chicago's underworld king. However he was never convicted of any of the deaths.

Keeping the war between the two Chicago area gangs burning hot, Capone ordered the death of Earl "Hymie" Weiss who had taken control of the North Siders.

In October of 1926, Weiss and four companions were walking down the street, ironically just feet from O'Banion's flower shop, when machine gun fire rang out in ambush leaving Weiss and one of his bodyguards dead and his lawyer wounded.

Following Weiss' death, a twenty seven year-old mobster named Vincent Drucci rose to power. He was unlike any of his predecessors; he was cocky, young, good looking, flashy and loved to be in the spotlight. In 1927, he was picked up by police on a misdemeanor charge. As the police were taking him to be arraigned, one of the cops, probably on the Capone payroll and tired of Drucci's constant berating of the police, shot him several times while he sat in the back of the police car.

Drucci's death lead to the hard hitting George "Bugs" Moran ascending to the head of the gang and Moran had a deep hatred of Al Capone and vowed to see him dead.

In revenge, the North Siders gunned down some South Side men and Al "Scarface" Capone was determined to avenge the murders.

A scheme was concocted to lure "Bugs" Moran and some of his henchmen to a warehouse where the North Side gang stored whiskey.

One of Capone's men got someone close to Moran to tell him that a shipment of hijacked whiskey would be delivered to the warehouse on the morning of February 14, and that Moran should be there.

Seven of Moran's men waited in the warehouse for the delivery. Moran was late in getting there and just as he was driving up, he saw a police car pull up to the warehouse. He drove on by and stopped down the block for a coffee and to wait for the police to leave.

Two uniformed cops and three plain clothes men walked into the warehouse. A short time later the repeated blasts of "Tommy guns" rang out shattering the silence of the cold February morning.

When the police arrived they found that Moran's men had been lined up along the wall and machine gunned down. The blood ran thick on the floor of the warehouse, seven men had mercilessly been shot, some almost cut in half by the high powered Thompson sub machine guns.

Six of the seven men killed in the massacre were hoodlums with thick files at the Chicago Police Station, one was a man without a criminal record.

The men who were killed in the warehouse were:

Dr. Reinhart H. Schwimmer: An optometrist who enjoyed associating with the gangster element. He was known to brag about his gangster friends. Dr. Schwimmer and his curiosity about the gangster life took him to the wrong place that night.

Peter Gusenberg: A character with a twenty-seven year history of crime and a leader in the Moran gang. He served time in Joliet Prison on burglary charges and was later sent to Leavenworth for robbery.

Frank Gusenberg: Brother to Peter, and a man with a violent past having served several jail and prison sentences.

Albert R. Weishank: Owner of the Alcazar Club and official of the Central Cleaners and Dyers company. The police theorized he had joined the gang just months earlier when Moran tried to take over the leadership of the cleaning and dying industry in Chicago.

James Clark: A brother-in-law to "Bugs" Moran and a career criminal having served four jail sentences for robbery and burglary.

John May: A known safecracker with a jail record who worked for the North Side Gang.

Adam Heyer: A criminal with four visits to jail for robbery and running a confidence game and owner of the S. M. C. Cartiledge Company where the murders took place.

The February 14th attack on the seven men in a "Bugs" Moran liquor warehouse became known as the "St. Valentines Day Massacre." The results of the massacre backfired on Al Capone. His power seemed to dim after the highly publicized killings. Chicagoans were disgusted with the brutality of the killings and the tide began to switch from protecting and respecting the gangsters to a desire by the public for peace and safety on the streets.

The federal government also was fed up with the wholesale murder and lawlessness of Chicago which spread its tentacles across the nation. The new President, Herbert Hoover, was determined to rid the nation of the scourge of hoodlums who held some of the country's greatest cities hostage. Hoover assigned U.S. District Attorney George E. Johnson to investigate and put an end to Al Capone's violations of the Volstead Act which made it illegal to manufacture, sell or consume alcohol. He also

A young Prohibition Agent, Eliot Ness, was appointed as a special agent to put an end to the bloodshed and tyranny of Al Capone. Photograph from the FBI Collections.

Eliot Ness and his Prohibition men raided Capone's breweries disrupting his criminal empire.

ordered Andrew Mellon, the Secretary of the Treasury, to search out ways to get at Capone through income tax evasion.

Johnson selected Eliot Ness, a young Prohibition Enforcement officer known for his high integrity, honesty and fearlessness, to lead the fight against Al Capone's bootlegging operations. Ness soon gained notoriety throughout the country for his high profile raids on the Capone breweries and distilleries. On one occasion, 18 distilleries were raided and shut down in one night. The equipment was put into storage for evidence and fifty two people were put into jail.

Ness then turned his attention towards Capone's breweries. He raided several of them, arresting more of Capone's men, but more importantly, the raids shut down operations that produced hundreds of gallons of beer a day, worth tens of thousands of dollars. One of Capone's main sources of money was drying up.

On thc other front, the Department of Treasury was working to put Al Capone away on charges of tax evasion. Al Capone lived a lavish lifestyle without any means of income. It used to be that a gangster could hide from paying taxes because they were protected by the constitutional right of freedom of self-incrimination. A criminal could avoid paying taxes on money illegally gained and not be found guilty of tax evasion because by paying it he would incriminate himself.

The Sullivan decision handed down by the Supreme Court threw that argument out and opened up the charges of tax evasion to be pressed against anyone who could not prove income to justify the money they spent.

It is said that Al Capone earned $100,000,000.00 per year on all of his illegal operations but paid zero in income taxes. He had several cars, homes in Illinois and Florida, office buildings and spent money freely. But he had no legally visible means of support. This is what was Al Capone's downfall. He was arrested and found guilty of tax evasion and, on October 17, 1931, was sentenced to eleven years in jail. Al Capone never lived long enough to serve out his jail term much less to be released back into society. On January 25, 1947, Alphonse Gabriel Capone died of complications caused by syphilis.

ROCCO PERRI, CANADA'S "KING OF THE BOOTLEGGERS"

Rocco Perri was born in Italy and immigrated to the United States to find the promised land. Growing up in the slums of New York, Rocco learned from his humble beginnings to take what he could and to despise authority.

Rocco moved to Canada and bounced around Southern Ontario working any unskilled job he could find. It was in Toronto in 1912 where he found the love of his life, Bessie Starkman. The fact that she was already married with two young children didn't prevent the two from leaving and starting a life together in St. Catherine with him working as a laborer on the Welland Canal.

When World War I began, the work on the canal was halted and Rocco was out of work. The couple moved to Hamilton and he found work as a salesman for a macaroni company. But when the Ontario Temperance Act was passed, Rocco and Bessie saw a way to make some real money.

The Act made it illegal to sell alcoholic beverages in Canada. Liquor could still be legally made and legally consumed, it just couldn't be sold within the boarders of Ontario, Canada, to Canadian citizens. This protected the Canadian liquor distilleries, and allowed companies like Hiram Walker, Seagram's, and Gooderham and Wort to still sell their product in other Canadian providences and other countries.

The Perri's purchased a grocery store which sold liquor. To maximize their profits they cut the quality Canadian whiskey with moonshine turning one bottle of legitimate whiskey into several altered bottles. Rocco was caught by the authorities and paid a hefty fine but he learned a valuable lesson that would be helpful for his next venture in the black market alcohol trade. The lesson: don't do the dirty work yourself; hire someone to do it for you.

When the 18th Amendment to the constitution of the United States became law, Rocco and Bessie were ready to supply the thirsty Americans with fine Canadian beer and liquor.

Rocco obtained a fleet of boats and trucks and several trusted employees to transport his products across Lake Ontario to U.S. cities. It wasn't long before the Perris had a reputation for being able to supply large amounts of beer and booze, a talent that did not go un-noticed by American crime bosses.

Rocco soon had contracts to provide booze to many of the mobsters in major cities of the United States. Perri's boats with cargos of contraband crossed Lake Ontario to Buffalo, New York. Some boats crossed the Niagara River at remote areas to the States but the majority of Perri's illegal beverages were trucked across Ontario to Windsor where it crossed the Detroit River destined to be sold in speakeasies in Detroit, Chicago, St. Louis and other thirsty American towns. Rocco Perri, never shy, proclaimed himself the "King of the Bootleggers!"

Rocco Perri negotiated the deals and arranged deliveries while Bessie took care of the financial end of their illegal enterprise. The couple's empire, which included working with many criminal organizations throughout Canada and the United States, was built on the dead bodies of many. Gangland wars were common as bootleggers encroached on one another's territories. Mobsters were shot in broad daylight on city streets, vehicles were bombed and some gangsters disappeared forever probably in concrete shoes at the bottom of Lake Ontario, while others turned up as floating bloated bodies. The Perris' were never indicted in any of the murders but they were suspected of involvement in murders on both sides of the border.

Rocco and Bessie's criminal empire was shattered on August 13, 1930, when hit men hiding in the Perri garage opened fire and killed Bessie. A distraught Rocco gave Bessie a lavish funeral fitting royalty and vowed to kill those who murdered his beloved wife.

Rocco continued in his criminal ways. After the repeal of prohibition, he expanded his involvement in the narcotic trade. Through his connections with criminals in both Canada and the United States, he smuggled drugs such as morphine, cocaine and heroin into the States

Fourteen years after Bessie met with death at the hand of an assassin, Rocco died in a manner befitting the gangster he was; he went out for a walk but never returned. He just disappeared as so many mobsters had before him.

TOLEDO'S LICAVOLI BROTHERS

Cities all across America and Canada had their own mobsters running illegal activities. In Toledo, Ohio it was the Licavoli's who ruled the underworld.

Brothers Thomas (Yonnie) and Peter and their cousin James were born in Sicily and emigrated to St. Louis, Missouri. They lived in the slums but their parents held out hope that they would become priests or doctors. The streets won out and the boys turned to a life of crime.

They began working for the Egan Rats, St. Louis's most violent and ruthless gang. During prohibition, the Licavoli's saw the potential of big profits to be made in bootlegging and moved to Detroit.

With their brutality and brains, the Licavoli's soon were some of the most powerful mobsters in the Detroit area. They operated out of Wyandotte, a downriver suburb. The Licavoli's successes did not go unnoticed by Detroit's Purple Gang. Rather than fight, the two groups worked together to control the illegal alcohol, gambling, drug and prostitution business in most of Detroit and southwestern Michigan.

Yonnie and Pete decided to expand their operation and take over the illegal operation in Toledo, Ohio. But Jackie Kennedy, a young Irish bootlegger, was the owner of a very popular speakeasy in Toledo. He controlled the gambling and booze in the city and wasn't going to take the intrusion into his territory lying down. He was determined to protect his territory and drive off the Licavoli's.

The Licavoli's made an attempt on the life of Jack Kennedy in December of 1932. Kennedy and his girlfriend, Louise Bell, were driving in his car. Jack bent forward, reaching for something just as a burst of machine gun fire blasted from a passing car. The bullets whizzed by Jack's head but struck Louise, killing her instantly.

Jack refused to bow to the thugs from Detroit and continued bringing booze into Toledo. While Jack Kennedy was no better than the common criminal, in Toledo he was a popular criminal. A few months after Louise Bell was machined gunned to death, Jack was killed. He was walking on a Toledo beach when, on July 7, 1933, he was shot several times in his head and body.

The brutal and vicious death of the popular Toledo club owner resulted in the citizenry of Toledo crying out for the arrest and conviction of the killers. The prosecuting attorney made the case against Yonnie Licavoli and some of his men and they were quickly tried and convicted. Yonnie was sentenced to life in an Ohio prison ending the Licavoli siege on Toledo.

MINNESOTA'S ISADORE BLUMENFELD

Minnesota's contribution to this criminal collection is the infamous Isadore Blumenfeld, a.k.a. Kid Cann.

Kidd Cann, born in Romania to poor Jewish parents, migrated to Minneapolis, Minnesota in 1902. As youngsters, Isadore and his brother Harry became involved in petty crimes and small scale extortion. When the 18th Amendment became the law of the land, the Blumenfelds saw it as an opportunity to cash in on the unrelenting desire of the American public to drink alcoholic beverages.

Blumenfeld used strong arm tactics to get whatever he wanted and he was not above killing anyone who got in his way. His ruthlessness was legendary.

The Minneapolis brothers arranged the importing of spirits from Canadian breweries and distilleries and took delivery of the illegal cargo on the wharfs of Duluth, small fishing villages, and desolate coves along Lake Superior. They also made frequent trips to Stearns County, Minnesota, to purchase the quality whiskey, "Minnesota 13", and sell it on the streets of Minneapolis and other American cities. They had contracts to supply the quality whiskey all around the Midwest.

The brothers operated a legitimate perfume business and imported barrels of industrial alcohol presumably for use in producing their fragrances. While some alcohol might have been used in the factory, most of the alcohol was transported to backwoods stills. The inexpensive industrial alcohol was mixed with the booze distilled to increase the volume and to give it a little extra kick.

Kidd Cann and his brother expanded their criminal enterprises to include prostitution, gambling and extortion, in addition to their lucrative bootlegging operation. Their success in illegal enterprises had been credited to the amount of people they had on their payroll. Patrolmen, police chiefs, judges, prohibition agents, the mayor, city council and even the governor were reputed to be in the back pocket of the Blumenfeld's.

Isadore was indicted several times on a variety of charges but he was never convicted. Perhaps he was innocent of the charges or possibly it was his skill at bribing witnesses and jurors that got him off.

Kid Cann could be especially vicious. If you crossed him, you could meet your maker in short order. He was suspected in the deaths of many, including newspaper reporters who dared to tie him to beatings, deaths and other illegal activities.

After the 21st Amendment repealed prohibition of the manufacture, sale and consumption of alcoholic beverages, the Blumenfeld brothers continued to rule the criminal underworld of Minneapolis. Kidd Cann dominated the rackets and was suspected in other killings but he was not arrested for them until in 1961 when he was convicted of bribing a juror and violating the Mann Act when he transported a woman across state borders for reasons of prostitution.

Isadore Blumenfeld was sentenced to Leavenworth for eight years but was paroled after serving just three years. He "retired" to Miami, Florida where it was reported that he and several other retired gangsters dabbled in illegal real estate deals and questionable stock market transactions.

Bibliography

History Of The United States Coast Guard

Canny, Donald, I, "Rum War: The U.S. Coast Guard And Prohibition", United States Coast Guard Historian's Office.

Peterson, William D., "Images Of America; United States Life-Saving Service In Michigan" Arcadia Publishing, Chicago, Illinois, 2000.

Shanks, Ralph, Wick York, Lisa Woo Shanks, "The U.S. Life-Saving Service: Heroes, Rescue And Architecture Of The Early Coast Guard" Costano Books, Petaluma, California, 2000.

Stonehouse, Frederick, "Wreck Ashore", Lake Superior Press, Duluth, Minnesota, 1994.

The United States Coast Guard Historian's Office, http://www.uscg.mil/history.

The Wreck Of The *Henry Cort*

Michigan Shipwreck Research Associates, http://www.michiganshipwreck.org/cort.html.

The Muskegon County Chronicle, Muskegon, Michigan, December 1, morning and afternoon editions 1934.

The Muskegon County Chronicle, Muskegon, Michigan, December 3, 1934.

Billy Gow: Hero Of The Tug *Reliance*

Historical Collection of the Great Lakes, Bowling Green State University.

Swayze, David D., "Shipwreck!", Harbor House Publishers, Inc., Boyne City, Michigan, 1992.

Marquette Mining Journal, Marquette, Michigan, December 18, 1922.

The Evening News, Sault Saint Marie, Michigan, December 16, 18, 19, 20, 1922.

Heroes Of The Armistice Day Storm Of 1940

The Muskegon Chronicle, Muskegon, Michigan, September 12, 13, 14, 1940.

Shanks, Ralph, Wick York, Lisa Woo Shanks, "The U.S. Life-Saving Service: Heroes, Rescues And Architecture Of The Early Coast Guard", Costano Books, Petaluma, CA, 2000.

Stonehouse, Frederick, "Wreck Ashore U.S. Life-Saving Service: Legendary Heroes Of The Great Lakes", Lake Superior Port Cities, Inc., Duluth, Minnesota, 1994.

Swazey, David D., "Shipwreck! A Comprehensive Directory Of Over 3,700 Shipwrecks On The Great Lakes", Harbor House Publishers, Boyne City, Michigan, 1992.

"So Long Boys, And Good Luck."

Http://www.pasty.com/~barbspage/steinbrenner.html.

Marquette Mining Journal, Marquette, Michigan, May 11, 12, 13, 1953.

The Daily Mining Gazette, Houghton, Michigan, May, 11, 12, 13, 1953.

The Evening News, Sault Ste. Marie, Michigan, May, 11, 12, 13, 15, 1953.

United States Coast Guard Accident Report, Washington D.C., July 10, 1954, *Henry Steinbrenner* - a-9 Bd.

The Storm Of September 1930

Grand Valley Daily Tribune, Grand Haven, Michigan, September 27, 28, October 8, 1930.

Ludington Daily News, Ludington, Michigan, September 29, 1930.

Ludington Sunday Morning News, Ludington, Michigan, September 28, 1930.

Muskegon Chronicle, Muskegon, Michigan, September 27, 28, 1930.

The Sand Beach Life-Saving Crew And The Wreck Of The Schooner *St. Clair*

The Huron Times, Sand Beach, Michigan, October 5, 1888.

Quay, Charles, "The Wreck Of The *Mattawan*."

Monument in the Port Sanilac Cemetery, written be Robert Smith a member of the Sand Beach Life-Saving crew who made the heroic voyage from Sand Beach to Port Sanilac.

Captain Mattison And *D.L. Filer*

Historical Collections of Ohio, Diaries of S. J. Kelly, Plains Dealer, by Darlene E. Kelly.

Http://www..rootsweb.com/~usgenweb/oh/newspapers/erie/part 5.txt.

Http://www..rootsweb.com/~usgenweb/oh/newspapers/erie/part 5.txt.

The Detroit Free Press, Detroit, Michigan. October 21, 22, 1916.

Wachter, Georgann and Mike, http://www.vaxxine.com/advtech/-shipwrecks/sw2006/g&mw.htm.

Abigail Becker, The Heroine Of Long Point

Swazey, David D., "Shipwreck! A Comprehensive Directory Of Over 3,700 Shipwrecks On The Great Lakes", Harbor House Publishers, Boyne City, Michigan, 1992.

Abigail Becker: The Heroine Of Long Point, Harry Barrett's Lore & Legends, http://www.kwic.com/~pagodavista/abigail.html.

Abigail Becker, http://www.canadiangenealogy.net/heroine/abigail_beck-er.htm.

Abigail Becker, The Heroine of Long Point Canada, Robert B. Townsend, "http://bobsnautical.50megs.com/photo6.html"

Long Point, Ontario, *Wikipedia*, the free encyclopedia, http://en.wikipedia.org/wiki/long_point,_Ontario.

Right Time, Right Place To Be A Hero

The Port Huron Times Herald, July 17, 18, 1945.

Historical Collection of the Great Lakes, Bowling Green State University.

Swazey, David D., "Shipwreck! A Comprehensive Directory Of Over 3,700 Shipwrecks On The Great Lakes", Harbor House Publishers, Boyne City, Michigan, 1992.

Keeper Captain Kiah And His Heroic Surfmen

Burley, Dale, The Harbor Beach Catalog, 1982, Harbor Beach, Michigan.

Swazey, David D., "Shipwreck! A comprehensive Directory Of Over 3,700 Shipwrecks On The Great Lakes", Harbor House Publishers, Boyne City, Michigan, 1992.

The Huron County Tribune, April 1880.

The Point Aux Barques Lighthouse Preservation Society.

The Coast Guard Medal

Detroit News, Detroit, Michigan, May 22, 1994.

The United States Coast Guard Historian's web site.

The King And The Great Lakes

Http:/www.globalsecurity.org/military/systems/ship/steam5.htm.

Http:/www.terrypepper.com/lights'closeups/strang/htm.

State of Michigan Archives, Lansing, Michigan.

Detroit Daily Press, Detroit, Michigan, July 10, 1851.

Pirates On The Great Lakes

Brumwell, Jill Lowe, Drummond Island History, Folklore, and Early People, Black Bear Press, 2003.

Timeline, A brief History of the Fur Trade, http://www.whiteoak.org/learning/timeline.htm.

Wikopedia, the free encyclopedia, Colby Pirates.

"Michigan's Only Pirate", Criminal Justice resources, Michigan State Universities Library, http://www.lib.msu,edu/harris23/crimjust/history.htm.

"Dan Seavey: Pirate of the Great Lakes", Joe Ahlers, http://www.mikeonline.com/story.asp?id=141672.html.

"Local Pirate leave legacy of tale" Leslie Allen, *Marquette Monthly*, October, 2008.

French and Indian War, Ohio History Central, a product of the Historical Society, http://www.ohiohistorycentral.org/entry.php?rec=498.

Pirate Bill, Canada's Real Live Pirate, Robert Townsend, http://www.bobsnautical.50megs.com/photo5.html.

A brief history of the French & Indian War, The Philadelphia Print Shop, Ltd. http://www.philadelphiaprintshop.com/frchintex.html.

Rock Island Lighthouse Keepers, "Pirate William Johnson, 1853-1861", Mark A. Wenting, http://www.rockislandlighthouse.org/johnson.html.

Pirates And Lake Erie's Johnson's Island

Harper's News Monthly Magazine, "The Wine Islands Of Lake Erie" by Constance F. Woolson, Volume 47, Number 277, June 1873.

The New York Times, August 11, 1895.

Bibliography

The Prohibition Era: 1920-1933

The New York Times, New York, New York, October 1, 1922.

"The Michigan Whisky Rebellion"' Allen May, "Crime Magazine and Encyclopedia of Crime", http://www.crimemagizine.com/michigan-whiskyrebellion.html.

The Chicago Tribune, Chicago, Illinois, February 23, 1920.

The New York Times, New York, New York, February 24, 1920.

The Detroit Free Press, Detroit, Michigan, January 1, 1917.

The Detroit Free Press, Detroit, Michigan, April 28, 1918, May 1, 1918.

The Detroit Free Press, Detroit, Michigan December 27, 1919. February 19, 21, 22, 1919.

The Times Herald, Port Huron, Michigan, February 20, 1919.

The Detroit News, Detroit, Michigan, April 30, 1918.

Gervais, C.H., "The Rumrunner: a Prohibition Scrapbook", Firefly Books, Thornhill, Ontario, 1980.

Hunt, C.W., "Whiskey and Ice: The saga of Ben Kerr, Canada's most daring Rumrunner", Dunduen Press, Toronto, 1995.

Lansing City Pulse, "How a mayor helped crush Lansing's spirits.", Robert Garrett, http://www.lansingcitypulse.com.

Steinke, Gord, "Mobsters & Rumrunners of Canada", Folklore Publishing, 2003.

Wikipedia, the free encyclopedia, "Rum-Running", http:/en.wikipedia.org/wiki/rum-running.

Colbourg and District History: Ben Kerr, created by lauraberg, http://www.cdhsarchives.org/wiki/tiki-index.php.page=ben=kerr.

Lawerance, Gordon, "*City of Dresden*… a Night the Whisky Ship Ran Aground" an edited edition of Harry Barrett's "Lore & Legends", http://www..mysteries-megasite.com/frames/ghostships-frame.html.

Swayze, David D., "Shipwreck!", Harbor House Publishers, Inc., Boyne City, Michigan, 1992.

Historical Collection of the Great Lakes, Bowling Green State University.

Huron County Tribune, Bad Axe, Michigan, February 25, 1927.

Huron County Tribune, Bad Axe, Michigan, April 28, May 5, 1922.

Huron County Tribune, Bad Axe, Michigan, August 24, 1928.

Port Huron Times Herald, Port Huron, Michigan, August 1928.

The Harbor Beach Times, Harbor Beach, Michigan August 1928.

"The Way It Was..." by Al Eicher, Lakeshore Guardian, Febuary 2005.

Yesteryears Edition Of The Huron Daily Tribune, Bad Axe, Michigan, March 15, 1996.

Chicago Sun-Times, Chicago, Illinois, August 27, 2008.

Crop Science Society of America, Madison, Wisconsin, May 31, 2007.

Melrose Beacon, "Minnesota 13 Brings Light To Legends", Herman J. Lensing, September 12, 2007.

Huron County Tribune, Bad Axe, Michigan, February 25, 1927.

Plains Feeder, Views from the Feedlot,
http://feedlot.blogspot.com/2008/05/minnesota-13.html.

Sun Tribune, Minneapolis - St. Paul, Minnesota, "An Idea that was all Wet", Bill Ward, October 27, 2007.

Wikipedia, The Free Encyclopedia, http://en.wikipedia.org/wiki/minnesota_13.

Http://info.detnews.com/redesign/history/story/historytemplet.cfm?id=8.

Criminals Of The Great Lakes

The Detroit News, Detroit, Michigan, detnews.com. "Detroit's Infamous Purple Gang", Paul R. Kavieff,

http://info.detnews.com/redesign/history/story/historytemplet.cfm?id=183.

My Jewish Learning.com, "Bootlegging, Fraud, and Murder by a gang of Detroit Jews" Rabbi David E. Lipman,
http://www.myjewishlearning.com/history_community/modern/modernsocialgangsters/purplegang.html.

TruTV,, "The Purple Gang", Mark Gibben,
http://www.trutv.com/library/crime/gangsters_outlaws/gang/purple/1.html.

The Purple Gang, *Wikipedia*, The Free Encyclopedia,
http://en.wikipedia.org/wiki_purple_gang.

The Detroit News, Detroit, Michigan, detnews.com, "The Purple Gangs Bloody Legacy", Susan Whitail.

The Buffalonian, "William J. Donovan (a.k.a. "Wild Bill") Takes on Buffalo," Steven Powell, http;\://www.buffalonian.com/history/article/1901-50/donovanschwabcase.html.

The Licavoli Gang, http://www.geocities.com/jiggs2000_us/licavoliy.html.

"Stone Marks Grave Of Woman Who Took Bootlegger's Bullet," Robin Erb, Toledo Blade, Toledo, Ohio.

"The Historic Old West End of Toledo," *Wikipedia*, the free Encylopedia.

Nash, Jay Robert, "World Encyclopedia of Organized Crime", Da Capo Press, New York, 1993.

Kid Cann, *Wikipedia*, the free encyclopedia, http://en.wikipedia.org/wiki/kid_cann.

Murder Incorporated, Isadore "Kid Cann" Blumenfeld, July 5, 2006, http://toughjews.blogspot.com/2006/07/isadore-kid-cann-blumenfeld.html.

"Minnesota's Most Notorious: The Legacy of Kid Cann", Frank Vascellaro, http://wcco.com/local/kid.cann.isadore.2.368394.html.

Kid Cann, St. Louis Park Historical Society, http://www.slhistory/org/history/kidcann.asp.html.

Glossary

Aft/After - Towards the rear end of a ship.

Beam - The largest width of a ship.

Barge - A barge is designed or rebuilt as a vessel to be pushed or pulled by another ship. The barge, usually without any means of propulsion, transports cargo on the Great Lakes and its harbors and tributaries.

Bilge - Lowest part of the interior of a ship's hull.

Bit - A wood or metal structural post on a tugboat to which the towline is attached.

Boiler - A metal container where water is heated to create steam to power the steam engines.

Bollard - A post or cast metal device on a dock or wharf. The mooring lines of a ship are secured to it to hold the ship to the dock or wharf.

Bootlegger - A person who smuggles liquor or beer against he laws of the country.

Bow - The front end of a ship. The port bow, starboard bow refers to the left and right of the forward sides of the vessel.

Breakers - The waves crashing on the shore or rocks.

Bridge - The pilothouse or wheelhouse. The location on the ship where the ship is steered.

Capsize - When a ship turns over; rolls over to the port or starboard.

Checking Speed - To reduce speed.

Chief Engineer - The crew member who is in charge of the engine and machinery of the ship.

Companionway, Passageway - A passage, corridor, or hallway on a ship.

Davits - The brackets which hold lifeboats on the ship and are used to raise or lower the lifeboats.

Deadlights - Portholes.

Deck Cargo - Cargo transported on the deck of a ship rather than in the hold below deck.

Distress Signal - A signal used to alert others of a disaster at sea and call for assistance.

Drydock - An area where a ship sails in and the water is pumped out. The drydock allows repairs to be made to the hull below the water line.

Firebox - The chamber in which a fire is built to heat the water in the boiler on a steam powered ship.

First Mate - Second in command of the ship behind the captain. To be a mate, the person must meet federal licensing requirements.

Fog Signal - A steam or compressed air powered whistle sounded in times of reduced visibility to notify other ships.

Fore - The forward or front of the ship.

Foredeck - A deck towards the bow of a ship.

Founder - To sink.

Freeboard - The part of a ship's hull above the surface of the water.

Freshening - The wind increasing; the wind is freshening.

Full Ahead - A ship moving at top speed in a forward direction.

Gale Force Wind - A strong wind from 32 to 63 miles-per-hour.

Gangway - A passageway in a side of a ship where passengers and cargo enter and depart the vessel.

Glass or Glasses - A telescope or binoculars.

Harbor of Refuge - A harbor, natural or man made, designed for ships to seek refuge in times of severe weather.

Hawser - A large rope used to tow or secure a ship.

Hold - An area on a ship below the deck where cargo is stowed.

Holed - A hull of a ship being punctured in a collision or by striking an object.

Hull - The body of the vessel.

Jacob's Ladder - A rope ladder.

Keel - A large wood or steel beam running bow to stern at the lowest point of the ship. It forms the backbone of the ship.

Knot - Speed of a ship which equals one nautical mile per hour.

Lee - Protection; the side of a ship protected from the wind.

Lifeboat Station - The location of a lifeboat on a ship.

Life Jackets, Life Preserver - a personal floatation device used to support an individual at the surface of the water.

Lightened - Intentionally reducing the weight of the ship.

Lines - Ropes used on a ship.

List - When a ship leans to port or starboard, left or right, due to unevenly stored cargo or from taking on water.

Lumber Hooker - A ship involved with the transportation of wood products.

Mid Ship - In the middle of the ship from the bow to the stern.

Moonshine - Liquor made in unregulated stills. The homemade liquor was also known by various names, Corn Liquor, White Lightening, Skull Cracker, Ruckus Juice, Rotgut, Mule Kick and Hillbilly Pop.

Moonshiner - A person who made illegal liquor.

Mooring Lines - Lines used to secure a vessel to a dock, pier or wharf.

Passenger Ship - A ship designed to transport passengers on journeys extending more than just a few hours.

Range Lights - Lighted beacons spaced a distance from one another. A ship is on course when entering a river or harbor when it positions itself so the lights are aligned above one another.

Pilothouse - The Bridge of a ship; where the ship is controlled, steered, and navigated.

Pitch - The forward and aft rocking of a ship.

Port - The left side of a boat when looking towards the front.

Prohibition - The common term given to the thirteen years that the manufacture, sale and consumption of alcoholic beverages was not allowed in the Unites States.

Prow - The forward part of a vessel.

Radar - A device which sends out a strong beam of radio waves. When it strikes an object it reflects back. The radar can then determine the distance away, direction of movement and speed of the object.

Reef - A rocky or sandy feature at or near the surface of the water.

River Hogs - The men working in the woods during Michigan's white Pine era that floated the logs cut deep in the forests down stream to the saw mills.

Roll - The sideward, port to starboard, movement of a ship.

Schooner - A sailing craft having two masts, one fore, the other aft.

Shanty Boys - The lumber men who felled the trees in the forests of Michigan.

Shipping Lane - Areas of a lake which ships must be traveling in a specific direction; up bound or down bound lanes.

Shoal Water - Shallow water.

Slow Astern - A ship operating at slow speed in a backward direction.

Soo Locks - Situated between Lakes Superior and Huron the locks raise ships 22-feet to the higher level of Lake Superior, or lower down- bound ships to the Lake Huron level.

Starboard - The right side of a boat when looking towards the bow.

Stern - The aft or rear of a boat.

Swing Bridge - A bridge which rotates from a central position. The bridge remains open for ship traffic to pass and closes only when it is needed. It is usually used for railroad bridges.

Tanker - A vessel specifically designed to transport liquid cargo, such as gasoline or other forms of oil.

Telegraph - A mechanical devise used to transmit instructions to and from the bridge of a ship to the engine room.

Teamsters - The men working in the woods who skidded the logs to the river banks.

Tender - A small boat which carries crew, passengers and supplies between a ship or lighthouse and shore.

Tommy Gun - The slang term for the Thompson Machine Gun.

Tug - The workhorse of the lakes. Tugs have been used to assist other ships to docks, or through congested waterways, tow barges, to transport cargo and passengers, for fishing, ice breaking, as a wrecking vessel and as a rescue vessel.

Turned Turtle - A common phrase to describe a ship which has been capsized and its bottom is above the surface.

Wharf - A structure built along a waterway where ships can tie up and discharge or take on cargo or passengers.

Wheelsman/Helmsman - The person who met the licensing restriction to be qualified to steer the ship.

Yawl - A small boat on a ship used by the crew to get to shore from an anchorage.

ACKNOWLEDGEMENTS

No historical endeavor can be accomplished without the assistance and aid of many. I want to thank all of those who contributed.

Bayliss Public Library and Susan James for assistance with research and access to their photographic collection.

Boatnerd.com, the people who respond to my questions.

Bowling Green State University, Historical Collections of the Great Lakes, Bowling Green State University, Robert Graham, Archivist.

Buffalo & Erie County Public Library.

Conneaut Public Library, Conneaut.

Hugh Clark Great Lakes Photographic Collection, Hugh Clark for his assistance in Canadian geography and access to his photographic collection.

DeFrain, Leonard, the Harbor Beach Historian.

Door County Maritime Museum and Lighthouse Preservation Society, Sturgeon Bay, Wisconsin.

Dossin Great Lakes Museum. John Polacsek, Curator of Marine History, Detroit, Michigan.

Gerow, Ed, for sharing his vast knowledge of the Great Lakes and ships.

Harbor Beach Public Library, Harbor Beach, Michigan. Vicki Mazure, Director.

Historical Museum of Bay County, Bay City, Michigan.

Library of Michigan and the State of Michigan Archives, Lansing, Michigan.

Lower Lakes Marine Historical Society, Buffalo, New York.

Klebba, Ron, The Highlander Sea Shipwright, for sharing his knowledge of boating and boat construction and sailing.

McDonald, David, Bad Axe County Historical Society.

McGreevy, Robert, a special thank you to Marine Artist Robert McGreevy for allowing the use of some of his paintings in this book, http://mcgreevy.com.

Main, Tom and Linda, Caseville, Michigan for their assistance in sailing terminology, technique and for sharing their knowledge of the Great Lakes.

Mehringer, Tom, Kerry Whipple, Charlie Unbehaun, and Captain Gary Venet for sharing their knowledge on the Great Lakes and S.C.U.B.A. diving experiences. I keep learning more from them than they will ever know.

Milwaukee Public Library, Suzette Lopez of the Great Lake Collection.

Morden, Charlie, thanks for sharing his Great Lakes historical information.

Point Aux Barques Lighthouse Society, Port Hope, Michigan.

Port Huron Museum, Port Huron Michigan, for her assistance.

Presque Isle Lighthouse.

St. Clair Public Library, Port Huron, Michigan.

The Grice House Museum, Harbor Beach, Michigan.

Toledo Lucas County Library, James Marshall, Director.

United States Coast Guard, Historian's Office, Christopher B. Havern, Historian.

University of Wisconsin-Superior, JDH Library, Laura Jacobs.

Williamson, K. Don, for his knowledge of Great Lakes boating.

Wicklund, Dick, Lake Lore Marine Society, for access to his personal photographic collection.

Wisconsin Maritime Historical Society, Catherine Sanders,
Milwaukee, Wisconsin.

About The Author

Photo by Karen Kadar

Geography has played an important part in shaping Wayne "Skip" Kadar's love of the Great Lakes. Throughout his life he has lived in the downriver area of Detroit, Marquette, Harbor Beach, and at the family cottage in Manistique, Michigan. Growing and living in these rich historic maritime areas has instilled in him a love of the Great Lakes and they're maritime past.

This love has taken him in many directions. He is a certified S.C.U.B.A. diver and avid boater, having owned most all types of boats. He is involved in lighthouse restoration, serving as the Vice President of the Harbor Beach Lighthouse Preservation Society, and a consultant in lighthouse restoration.

Mr. Kadar enjoys studying and researching Great Lakes maritime history and has made presentations on maritime history around the Great Lakes region.

An educator for thirty years, Mr. Kadar has retired after 15 years as a high school principal.

Skip lives on the shore of Lake Huron in Harbor Beach, Michigan with his wife, Karen. During the summer they can usually be found at the Harbor Beach Marina, on the family boat "Pirate's Lady" or at the lighthouse.

Other Wayne Kadar Titles
By Avery Color Studio:

Great Lakes Passenger Ship Disasters

Great Lakes Freighter, Tanker, & Tugboat Disasters

Strange & and Unusual Shipwrecks On The Great Lakes

Great Lakes Collisions, Wrecks & Disasters,
Ships 400 To 998 Feet